Managing Creativity in Organizations

Managing Creativity in Organizations

Critique and Practices

Alexander Styhre

and

Mats Sundgren

First published 2005 by
PALGRAVE MACMILLAN
Houndmills, Basingstoke, Hampshire RG21 6XS and
175 Fifth Avenue, New York, N.Y. 10010
Companies and representatives throughout the world

PALGRAVE MACMILLAN is the global academic imprint of the Palgrave Macmillan division of St. Martin's Press, LLC and of Palgrave Macmillan Ltd. Macmillan® is a registered trademark in the United States, United Kingdom and other countries. Palgrave is a registered trademark in the European Union and other countries.

ISBN-13: 978–1–4039–4768–0
ISBN-10: 1–4039–4768–6

This book is printed on paper suitable for recycling and made from fully managed and sustained forest sources.

A catalogue record for this book is available from the British Library.

Library of Congress Cataloging-in-Publication Data
Styhre, Alexander.
    Managing creativity in organizations : critique and practices / Alexander Styhre, Mats Sundgren.
        p.   cm.
    Includes bibliographical references and index.
    ISBN 1–4039–4768–6 (cloth)
    1. Creative ability in business—Management.   2. Technological innovations—Management.   I. Sundgren, Mats, 1959–   II. Title.
    HD53.S79 2005
    658.3′14—dc22   2005045253

10   9   8   7   6   5   4   3   2   1
14   13   12   11   10   09   08   07   06   05

Printed and bound in Great Britain by
Antony Rowe Ltd, Chippenham and Eastbourne

Contents

List of Tables and Figures

Foreword

One may argue that people often speak of creativity in mystical tones – as though it were a prize that is possessed by only a few. When creativity *is* discussed, it comes up in contexts outside work, while innovation is almost always discussed in work settings. So does an absence of creativity mean an absence of innovation? This and other issues captured our interest in creativity within organizations. In many ways, pharmaceutical R&D implies that creativity is important, which emphasizes the importance of understanding the period that precedes innovation within new drug development. So the *organizational creativity* concept becomes more relevant than mere creativity.

When the two of us started to work together at the beginning of the new millennium, Mats brought the lingering concern in the pharmaceutical for the declining return on investment in R&D in terms of new blockbuster drugs into discussion. According to Mats, several industry representatives had, in a variety of arenas, expressed their doubts and concerns about the future of the industry. Needless to say, such a vast problem or challenge may be approached in a number of different ways, but we thought it would be helpful for practice to treat the R&D challenge, especially in the case of the pharmaceutical industry, in terms of being a 'creativity problem'.

The engagement with the organizational creativity literature and the broader literature on creativity made us aware that the lack of, or consequences of creativity in practice were not the only problems, but the very theory or theories of creativity per se became an increasing source of reflection in our joint research. Not only is the literature on organizational creativity disjointed and disperse, but it is also, in many cases, failing to address a series of ontological, epistemological, and methodological problems arising from the idea, or belief, or even promise, that there is such a thing as organizational creativity.

However, even though this book is an attempt to address some of the concerns facing us and other organizational creativity researchers, we do not want to abandon the idea of creativity. Organizational creativity is a useful term that makes sense and helps individuals in both industry and academic circles address a series of strategic decisions and choices in organizations. But, in common with all theory, the idea of creativity needs to be thoughtfully reflected upon and examined in detail – in addition to standing the test of empirical investigation – in order to qualify as theory. In this respect, we follow Karl Popper's idea that a scientific theory needs to be capable of being tested – that is, to be corroborated against empirical material. This does not imply, however, that one should embark on some one-sided positivist research agenda, aiming to make empirical studies the single yardstick for what qualify as theory and what fails to do so. Consequently, the subtitle of this book is 'Critique and Practices', pointing to the need for both theoretical analysis and empirical studies when formulating a scientific theoretical framework. Expressed differently, this book is written with the ambition of both providing some thinking about the underlying assumptions, beliefs and ideologies in the creativity literature, but also demonstrating that the idea of creativity is a helpful tool when examining and understanding the most complex and complicated R&D activities taking place in the pharmaceutical industry under the banner of new drug development. When treating organizational creativity is a *tool*, something to be employed in practices, a form of *techne* or *phronesis*, one may take Osborne's (2003: 522) warning words into account: 'We should be suspicious of the idea of creativity when raised to the power of a doctrine or a morality.' In order to avoid such a position, organizational creativity *qua* concept deserves a proper critique in the Kantian sense – that is, a systematic examination of the various components of a theory and its subsequent empirical applicability. It is our hope that in this book we have at least managed to start such an analysis.

This book is by no means an effort solely by two singular individuals; rather, it is influenced, supported by, and made possible through the help of a number of individuals and organizations. This book, and its research, is the successful product of collaborative research between academia and industry, and several people at AstraZeneca supported the research and its development. So our thanks go to: Sverker Ljunghall for company sponsorship, support, and interest; Elof Dimenäs for supporting and contributing to a rewarding integration of professional work and research; Curt Bengtson, for unreserved enthusiasm and challenging dialogue about leadership and creativity; Barry Furr and Martin Nicklasson, for fruitful discussions. We are deeply grateful to all respondents, and colleagues within AstraZeneca in Sweden, UK and US; Arvid Carlsson, Uli Hacksell, and Gerth Wingårdh; and employees at Carlsson Research AB, ACADIA Pharmaceuticals, and Wingårdh Arkitektkontor AB – for contributing to this book by dedicating their precious time.

We would also like to thank our colleagues at the Department of Project Management at the Chalmers University of Technology and the Fenix Research Program at Chalmers, Stockholm School of Economics, and École des Mines de Paris with whom we shared some interesting discussions. A special thanks go to Hans Björkman who – whatever the time or location – was always a keen and patient conversant; Flemming Norrgren, Rami Shani, Niclas Adler, Hans Glise, Armand Hatchuel, Jan-Eric Gustafsson and Marcus Selart during various stages of research engaged us in stimulating and rewarding discussions. From the creativity research community we would also like to express gratitude to Tudor Rickards and Göran Ekvall for interest and support.

And finally, a number of anonymous reviewers have helped us to sort out some of our at times premature ideas and theoretical elaborations in papers submitted to journals. But, first and foremost, we would like to thank our families for being supportive in every sense of the term.

ALEXANDER STYHRE
MATS SUNDGREN

1
Introduction: The Absence of Creativity in Practice and Management Writing

Introduction

It is common to argue that we live today in a society characterized by the increasing influence and importance of intangible resources such as intellectual capital, know-how and knowledge. Since the 1970s, sociologists, economists and political analysts has debated the wide-ranging change from a society and an economy based on industrial production and manufacturing to one centred on service industries and intangible products and services. The movement from the primary sector of agriculture to the secondary sector of manufacturing was named the 'Industrial Revolution'. To date, the next movement from the secondary sector of the economy, manufacturing, into the third sector of service industry has not yet been given such a spectacular label, though the terms 'knowledge society' and 'information society' have been made popular; yet the changes can be argued to have been almost as influential. One of the most important implications to be drawn from this change is the emphasis on knowledge-intensive industries. Today, domains of the economy such as the finance, pharmaceuticals, and higher education sectors are playing increasingly important roles in western societies. These different industries share the common feature of being dependent on the use of intellectual capital and know-how. The emphasis on intellectual capital or knowledge as one of the major organizational resources has been one of the dominant traits of management studies since the mid- 1980s year. Despite this, creativity still stands outside the orthodoxy of management studies (Rickards, 1999). A substantial amount of business school research and organization theory has been dedicated to the study of how organizations and firms develop, share, and make use of know-how and expertise. This scientific inquiry has been pursued under such banners as knowledge management, organization learning, and innovation and R&D management. Other sub-disciplines within organization theory and management studies include research on entrepreneurship, the use of consultancy services in organizations, and strategic management theory. By and large, the ability of organizations and firms to make use of intellectual resources is portrayed as a source of sustainable competitive advantage. This book aims to address one of the key processes in exploiting intellectual resources, namely creativity, defined as the creation of new ideas, goods, and services through the exploitation of an existing stock of resources and know-how. Creativity remains one of the most contested, yet least understood processes or activities within this wide-ranging field of interrelated theories and practices making use of intellectual resources. First, the notion of creativity has connotations of somewhat mystified and mythological processes of creation among highly talented or/and specialized groups of individuals such as artists, film directors, and scientists. In this context, creativity is envisaged as some kind of 'divine breath' that flows through the selected few at a very specific moment of creation wherein extraordinary insights or deeds are enabled. This view of creativity has been widely exploited in popular culture in literature and films and in science mythology; the extraordinary writer or scientist's points of bifurcations and change of perspective remain one of the favourite mythemes in western culture. The history of science is filled with colourful stories of masterful men and women gaining insights in their laboratory work or even in their dreams. Think, for example, of the French chemist August Kekulé's dream of snakes biting one another's tails, constituting a chain reminding of the benzene rings Kekulé modelled on the image of the snakes, thereby advancing chemistry to a new level (Roberts, 1989). As a consequence of the emphasis on the residual explanations of creativity in popular culture, creativity has largely remained unproblematized and excluded from systematic reflection.

For the second, creativity has been treated as an *ex post facto* construct rather than a process that may be subject to systematic and thoughtful managerial practices. This suggests that creativity is based on a circular argument suggesting that creative people are people who have already proven to offer creative solutions. According to this account, creative solutions are then little more than those which, in hindsight, have proven to be successful. The inability to take an *ex ante* perspective on creativity has added further to the 'mystification' of creative processes because creativity has been treated as something that one cannot fully control. This is not to say that creative processes are removed from chance, local conditions, and other contingencies, but to suggest that creativity can actually be *managed* rather than being an effect of various conditions of which one can only control a subset. Taken together, the notion of creativity has yet to be accorded a proper role in management thinking. When the term is invoked, it is often in terms of a metaphor or poetic expression rather than as a rigorous scientific construct. In addition, the notion of creativity is surrounded by some degree of anxiety because it is generally conceived of something good – 'being creative' remain a positive marker – at the same time as there is little advice provided about how to become creative or manage creative processes or creative people. This anxiety has been exploited in numerous 'How-to' handbooks and self-help books, primarily slanted towards the psychology discipline, and all of which promise to provide guidelines on how to make use of one's creative faculties. When taking a critical perspective, one may refer to this body of literature – or at least some parts of it – as a form of *kitsch*. In the genre of management writing, the literature on creativity may be examined as a management fad and buzzword. Again, this is not to suggest that the notion of creativity is not helpful in management practice; it is rather to state that if the notion of creativity will achieve the status of a firm scientific construct, it needs to undergo the same critique and examination as other scientific concepts. Therefore, a critique of the notion of creativity and the practices of managing creativity remains an important part of scientific programme aiming to advance creativity as a key process in the emerging knowledge society. This book is an attempt to contribute to this research programme and is based on the following two propositions: (i) That creativity is, in theoretical terms, relatively little explored in relationship to for instance knowledge management and organization learning; and (ii) that the study of the practice of managing creativity is primarily based on either quantitative research or anecdotal evidence, thereby portraying creative work as something detached from its social settings and context, or something extraordinary that is complicated to transfer to new domains. As a consequence, there is a need for (i) a more elaborated and systematic critique of the constructs of creativity and management of creativity; and (ii) more detailed and contextualized case studies of how creative work is organized and managed in workplaces. Since creativity is here examined as what it is possible to manage and what is emerging in organized settings, we will speak of *organizational creativity* rather than simply creativity. The line of demarcation between these two forms of creativity is a fluid and permeable one, and serves primarily to distinguish between the romantic view of the creative individual – in most cases an artist or a scientist – and creativity as an organization resource that is employed when creating sustainable competitive advantages. In the text, we use the two terms interchangeably, but when we speak of 'creativity' we assume it is creativity that is subject to managerial practices and taking place within the organization. Since many of the writers that have contributed to the analysis of organizational creativity are actually speaking of 'creativity' rather than 'organizational creativity', we do not want to exclude those contributions on the basis of a semantic distinction. We therefore suggest the following tentative definition of organizational creativity in the context of pharmaceutical R&D, the principal domain of empirical investigation in this book:

> *A variety of activities in which new ideas and new ways of solving problems emerge through a collaborative effort by promoting dialogues that involve multiple domains of scientific knowledge to produce value for the organization's mission and market.*

This definition of organizational creativity can be seen as a synthesis of aspects taken from both academicians' and practitioners' perspectives. The practitioner's perspective has a more specific customer and market-driven focus, emphasizing dimensions of the actual work and of value (Andriopoulos, 2001). The academic perspective emphasizes aspects of novelty, diversity, and motivation and generally treats creativity as being an unbounded enterprise (Gioia, 1995). The two perspectives also reflect the idea that creativity in organizations involves divergent and convergent thought and action before it becomes effective. The definition also emphasizes that creativity is not solely about delivering new candidate drugs, but also includes all the activities in the pharmaceutical industry: new drug development activities, strategic management decisions, and human resource management practices (Jeffcut, 2000). So the definition of creativity encompasses the entire organization and consists of a multiplicity of activities.

Before moving on to the practical and theoretical positioning of the notion and idea of organizational creativity, we will further anchor the subsequent discussion in a broader social and managerial context in many cases referred to as the knowledge society and the knowledge-based firm.

The knowledge society and its consequences

By the end of the 1960s and at the beginning of the 1970s, many sociologists had begun to identify a general movement from the industrial to the post-industrial society. In 1973, the American sociologist Daniel Bell formulated his highly influential ideas of the emerging *post-industrial society* in which manufacturing industry accounts for a decreasing share of the value produced in industry and new forms of economic activities derived from the use of know-how and intellectual capital become increasingly prominent. Prior to that, the French sociologist Alain Touraine (1971) offered a similar, albeit more critical, account of this concept. For Touraine (1971: 12), the new society's distinguishing mark was that it was neither land not labour but knowledge that was now the primary production factor. According to Touraine, this emphasis on formal and systematic knowledge has far-reaching societal consequences. Firstly, it alters the class structure: '[t]he new dominant class is defined by knowledge and a certain level of education' (Touraine, 1971: 51). The middle class has always been dependent on education as its most important distinguishing feature (Ehrenreich, 1989; Bourdieu and Passeron, 1977). In the post-industrial knowledge society, this tendency is further accentuated. As a consequence, dominant classes, Touraine argues (1971: 61), 'dispose of knowledge and control *information*'. The access and control over information and knowledge is therefore what constitutes class and influence in the new society.

While the new social regimes sketched by sociologists such as Bell (1973) and Touraine (1971) were initially regarded as thoughtful reflections on a society undergoing substantial changes during the 1970s, it was not until much later that such ideas penetrated management literature more generally. The Japanese manufacturing industry – and primarily the automotive industry – started to pose a real threat to American companies at the end of the 1970s. As a consequence, in the 1980s, intangible organizational resources such as organization culture were explored as underlying factors explaining sustainable competitive advantage. In the 1990s, interest in management literature had turned to intellectual resources such as the organization's capacity for learning and changing and its use of its know-how and expertise. The following quote from Stewart's 1997 bestseller is representative of the message provided in the emerging literature on knowledge management:

> You win because today's economy is fundamentally different from yesterday's. We grew up in the industrial age. It's gone, supplanted by the information age. The economic world we are leaving was one whose main sources of wealth were physical. The things we bought and sold were, well, *things*; you could touch them, smell them, kick their tires, slam their doors and hear a satisfying thud. Land, natural

resources such as oil and ores and energy, and human and machine labor were the ingredients from which wealth was created. (Stewart, 1997: x)

Here, a new world order is being introduced. Stewart (1997) argues that while the physical resources used to be of central importance, today it is the knowledge-based resources that make the difference. This movement from manufacturing and the realm of the tangible to knowledge-based production and an increased reliance on intangible and intellectual resources had direct implications for management practice. Barley and Kunda (1992) examined the changes in what they refer to as *management ideology* over the period 1870–1992. In Barley and Kunda's account, management theory and practice have altered between *normative* and *rational* ideologies – that is, between managerial practices that, on the one hand, aim to establishing rational tools and techniques for the day-to-day management of operations, and, on the other, propose different normative statements on how management should be conducted. The period 1955–1980 was characterized by what Barkey and Kunda term 'systems rationalization': 'All systems rationalists regardless of discipline peddled programmatic techniques or universal principles that would enable managers to plan, forecast, and act more effectively. Accordingly, each camp draw moral, if not technical inspiration from scientific management' (Barley and Kunda, 1992: 379). After 1980, Barley and Kunda argue, organization culture – a normative ideology in their account – became the dominant research topic in management literature. After interest in organization culture had waned at the end of the 1980s and during the early 1990s, a variety of new management practices and concepts such as empowerment, projectification, and teamwork, were suggested as the key components of the new managerial system. In the emerging knowledge-based society, several contributors argued, managerial practices could not rely on outmoded Taylorist and Fordist management routines; rather, new practices had to be conceived of and become established in organizations. The dominant and recurrent theme in these new managerial practices was normative control rather than direct inspection and rational control. Rather than adhering to Frederick Winslow Taylor's bleak view of the co-worker, it was McGregor's (1960) so-called 'Theory Y model' of the co-worker – introduced in the management book *par préférénce* of the 1960s (Frank, 1997) – that served as the role model. In his acclaimed Theory Y ideal type, McGregor postulated among other things that 'the capacity to exercise a relatively high degree of imagination, ingenuity, and creativity in the solution of organizational problems is widely, not narrowly, distribution in the population'. Nothing could be farther from Taylor's talk about a co-worker being 'stupid as an ox' and the subsequent need for a substantial division of labour and the separation between 'brain work' and 'manual work'. The change from system rationalization to normative control implied new forms of management practice but also a change in focus on what are legitimate research questions in academic research on management practice. During the 1990s, a large number of books were published that further explored and developed ideas about the knowledge-based organization (Teece, 2000) or even the knowledge-based capitalism (Burton-Jones, 1999). The knowledge society had penetrated organizational lives.

Somewhat surprisingly, given the increased emphasis on the co-workers as intelligent and responsible human beings, relatively little is said about creativity in this wide-ranging discussion about the knowledge-based society. At best creativity was implied in the deployment of knowledge-based resources in organizations. Otherwise, there appeared to be a rather modest concern for the effects and importance of the co-workers' creativity. As we will see in the next section, organizational creativity may actually be a highly useful concept for addressing a number of pressing managerial concerns in, *inter alia*, the pharmaceutical industry.

The practical perspective: the absence of creativity

Our argument is that there is a need for a critical evaluation of the construct of creativity if it will hold water in studies of management practice in organizations. But what is the status of creativity in organizations and companies? Does the notion of creativity, to speak with Weick (1979), *make sense*? Is it a label that practicing managers find convenient to use when talking about certain activities and events in their day-to-day operations? Findings from empirical research suggests that is not the case. On the other hand, there are things occuring in organizations (events and occurrences, controversies and agreements) that practicing managers may consider to be moments of creativity, moments when new ideas and new images are formulated and jointly shared within a community of practice. Creativity is then a concept that might well play a role in the language-games of practicing managers.

If one takes a step back and assumes that what is referred to as creativity is capable of capturing some good productive activities in organizations in terms of the creation of new ideas and insights, then one may turn to, for instance, the pharmaceutical industry to examine what are the demands on the industry for long term sustainable performance.

Changes in the pharmaceutical industry and the need for creativity

By almost any measure, including R&D intensity and use of new scientific concepts, the pharmaceutical industry is a classic high-technology, science-based industry (Santos, 2003; Pisano, 1997). The industry shares many characteristics with other technology-intensive industries but also has some unique features, such as its highly regulated environment, long development cycles, and high-level risk and cost in the research process (Cardinal, 2001). In this context, the pharmaceutical industry depends on its leading-edge scientific capabilities and new scientific advances and technologies (Yeoh and Roth, 1999). Traditionally a relatively stable, conservative knowledge-based industry, the pharmaceutical industry now faces more intense competition from biotechnology firms that are part of new economic structures. The pharmaceutical industry has a long history of initial innovative breakthroughs (first-in-class) or paradigmatic innovation, followed by slower, stepwise improvements of such initial successes (best-in-class) or application-based and modification-based innovations (Hara, 2003; Horrobin, 2002; Achilladelis, 1999).[1]

Two important periods can be identified in the history of innovation within the pharmaceutical industry. In the first period – from 1820 to 1930 – scientific methods were adopted to purify diverse natural and synthetic materials, which generated clusters of pharmaceuticals (e.g., alkaloids, serums, antipyretics, analgesics, and hypnotics[2]). In the late nineteenth century, the industry was considered to be a specialized branch of the chemical industry. The second period, from about 1950 to the late 1980s, is often called the 'golden age' of pharmaceuticals (Lacetera and Orsenigo, 2001; Pisano, 1997). This period offered large R&D opportunities and unmet needs for pharmaceutical companies. As Pisano (1997: 55) points out:

---

[1] According to Hara (2003), innovation in the pharmaceutical industry can be divided into (i) paradigmatic, (ii) application, and (iii) modification-based innovation. Paradigmatic innovation occurs when neither the compound nor the application is known beforehand. Application innovation occurs when the compound is known but not the application. Modification-based innovation does not represent incremental innovation but refers more to what Kuhn (1970) calls normal science. Modification-based innovation is based on past scientific achievement and existing therapeutic approaches and does not challenge them.

[2] Examples of compound cases: morphine, salvarsan, quinine, cocaine, ether, and barbiturates.

Faced with such a target-rich environment but very little detailed knowledge of the biological underpinnings of specific diseases, pharmaceutical companies invented an approach called random screening. Under this approach, natural and chemically derived compounds were randomly screened in animal models for potential therapeutic activities.

During this period, serendipity played a key role. In fact, it was not uncommon for companies to discover a drug to treat one disease, while searching for another (Pisano, 1997).

During this period, pharmaceuticals became a truly research-intensive industry, which generated several radical generations of innovations (e.g., antihistamines, antibiotics, corticosteroid hormones, beta-blockers, and antihypertensive drugs[3]). These innovations revolutionized the structure and business practices of the industry (Achilladelis, 1998). Throughout its history, the industry has maintained a close and fruitful relationship with institutions of academic research in chemistry, medicine, and life sciences.

Since the 1970s, some pharmaceutical firms have enlarged to become enterprises, comparable in size to those found in the electronics, telecom, or automotive industries. But now the industry finds itself facing crucial choices in a difficult economic and regulatory environment (Drews, 2003). During the last 20 years, the industry has undergone radical transformation and consolidation. One of the most serious problems the industry faces is rapidly increasing R&D costs, coupled with relatively small increases in the output of new products. Since the early 1990s, the industry has had to deal with new economic and technological changes (Hullman, 2000). Regulatory demands have become significantly stricter in the last decade, and this has resulted in less product exclusivity and price flexibility. New patent regulations (Waxman Hatch Act, 1984) allow companies to launch generic versions of drugs that have gone off patent without having to undergo extensive clinical trials, resulting in intense, generic competition (Pisano, 1997). Another change is increased drug research costs, including extensive clinical programmes that often involve more than 20,000 patients in the later development phases (Zivin, 2000). A third issue is that many

---

[3] Examples of compound cases: penicillin, cortisone, beta-blockers, calcium antagonists, and ibuprofen.

successful products developed during the 1980s are now going off patent. So the industry tendency is now to maintain high focus on decreasing time to market and reducing bottlenecks to optimize the patent term of the product (Tranter, 2000; Drews, 1997).

Another important change since the mid- 1990s is the unprecedented rate of development in computer science and discoveries in other scientific domains, such as biotechnology (e.g., genomics, proteomics, and bioinformatics).[4] These have resulted in greater competition and radical change in the pharmaceutical discovery process (Jain, 2000). This process has moved from a more classical random screening approach, towards a rational drug design that is based more on detailed knowledge and involves sophisticated technologies, such as computer-aided drug design (CADD), combinatorial chemistry (CC), high-throughput screening (HTS), and genetic engineering – in search of increasingly complex and potentially more effective molecular structures as bases for new products. In more recent times, there has been a critical debate as to whether the industry has put too much effort into, and focused too exclusively on, these technologies (e.g., Horrobin, 2003; Reiss and Hinze, 2000).

For large pharmaceutical companies, the main issue is to sustain average industry growth. For every 1–1.5 per cent share a company has of the world market, one new product must be introduced each year, which will sell for more than US$ 400 million annually (Horrobin, 2000). This need to optimize revenues has resulted in the consolidation of many pharmaceutical companies through a process of mergers or acquisitions. Table 1.1 illustrates the trend to create huge R&D organizations.

Unpredictability in the research process is also an important issue. Most pharmaceutical companies have experienced high project attrition: on average, only one per cent of early discovery projects end up as products in the market (Dohlsten, 2003), which is understandable when considering that the number of mergers, or acquisitions, during the last decade has been significant. As Table 1.1 shows, huge pharmaceutical companies like

---

[4] Genomics is the large-scale use of small molecules to study the function of gene products. Proteomics, a branch of functional genomics, is the large-scale analysis of polypeptides during cell life; its purposes are to catalogue proteins that our genes encode and to decipher how these proteins function to direct the behaviour of a cell or an organ. Bioinformatics is a cross-discipline of computer science and biology; it seeks to make sense of information from the human genome, to find better drug targets earlier in drug development (Hopkin, 2001; Howard, 2000).

AstraZeneca[5] have been created since the mid- 1980s. For example, both
GlaxoSmithKline and Pfizer are the results of three or four mergers. Consolidation continues to be a major event in the pharmaceutical industry,
mainly because of the effects of innovation deficit (Drews, 1998, 2003).

Table 1.1   Selected international mergers and acquisitions in the pharmaceutical
industry[6]

Date	Company	Company	New company
1985	Searle	Monsanto	Monsanto
1989	Squibb	Bristol-Myers	Bristol-Myers Squibb
1989	SmithKline-French	Beecham Group	SmithKline Beecham
1994	American Cyanamid	American Home Products	American Home Products
1994	HCA-Hospital	Columbia Healthcare	Columbia Healthcare
1995	Upjohn	Pharmacia	Pharmacia-Upjohn
1996	US Healthcare	Aetna Life & Casualty	Aetna Life & Casualty
1996	Sandoz	Ciba-Geigy	Novartis
1999	Monsanto	Pharmacia-Upjohn	Pharmacia Corp.
1999	Zeneca	Astra	AstraZeneca
1999	Rhône-Poulenc	Hoechst	Aventis
2000	Warner-Lambert	Pfizer	Pfizer
2000	SmithKline	Glaxo-Wellcome	GSK
2001	Dupont	Bristol-Myers Squibb	Bristol-Myers Squibb
2002	Immunex	Amgen	Amgen
2002	Takeda	Gruenenthal	TakedaGruenenthal
2003	Pharmacia Corp.	Pfizer	Pfizer
2004	Aventis	Sanofi-Synthelabo	Sanofi Aventis

[5] AstraZeneca is a major international healthcare company engaged in the
research, development, manufacture and marketing of prescription pharmaceuticals and the supply of healthcare services. It is one of the world's leading pharmaceutical companies with healthcare sales of over $18.8 billion in 2003. The
company operates within seven therapeutics areas: neuroscience (CNS & pain
control), cardiovascular, gastrointestinal, oncology and infection, respiratory
and inflammation. AstraZeneca is ranked number five in the industry for R&D
expenditure (US$3.5 billion in 2003) and for employees in R&D (more than
11,000). The head office is located in London and the company have seven sites
in Sweden, the UK and the US.
[6] *The Economist* (1998) 'The mother of all mergers', 5 February issue, and
www.drugintel.com.

One may argue that the industry has failed to radically adapt to new changes and thus to balance sales and R&D costs (Dimenäs *et al.*, 2000).

All of the 10 largest companies now have R&D organizations with between 6,000 and 10,000 researchers and with R&D budgets of between US$3 and US$5 billion[7]. The recent acquisition of Pharmacia will make Pfizer the world's largest pharmaceutical company by far, with combined sales of around US$48 billion, 30,000 sales representatives worldwide, and an annual R&D budget of more than US$7 billion. Pfizer will also become the first company in recent history to possess more than a 10 per cent of the global drug market.[8]

The research strategy adopted in many of these companies has thus come to focus on products that are expected to become mega brands with anticipated revenues of more than US$1 billion annually. As a result, the total worldwide new medical entities (NMEs) launched annually have fallen every year from 80–100 in the 1960s, to 50–60 in the early 1980s, and 30–40 in the late 1990s (Horrobin, 2000). Although major drug companies spend more than US$30 billion annually on R&D, this trend continues (see Figure 1.1).

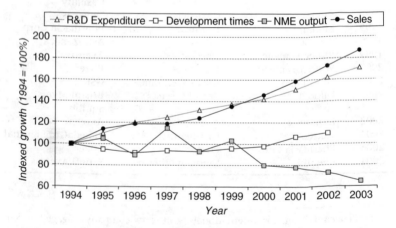

Figure 1.1   The pharmaceutical R&D productivity challenge[9]

---

[7] *The Economist* (2002) 'Mercky prospects', 15 July issue.
[8] *The Economist* (2002) 'Mating and waiting', 20 September issue.
[9] *The Economist* (2003) 'Big trouble for big pharma', 4 December issue, and CMR International. Includes data from pharmaceutical companies of all sizes; CMR solicits data from > 60 companies.

Consequences and implications

The focus on blockbuster drugs in large pharmaceutical companies, the significant number of mergers, acquisitions, and joint ventures, and the increasing emphasis on shareholder value and stock market performance – combined with a stronger regulatory environment – have essentially reduced the degree of freedom in pharmaceutical practice (see, for example, and Schmid and Smith, 2002a). Among several consequences for the industry, large multinational firms have tended to become increasingly process-oriented and to rely on standard operating procedures and other forms of bureaucratic routine and standardization (Hullman, 2000). The pharmaceutical industry is becoming bureaucratized, and a dominant strategy in such companies has been to streamline the research process (Horrobin, 2000). Rigorous project and portfolio management is implemented in an attempt to increase the number of products launches and to reduce R&D costs (Johnson, 1996). In many cases, this results in a flurry of benchmarking initiatives and the formulation of overly ambitious business goals (Halliday, 1999; Hughes, 1998).

Moreover, the overall industry has tended to become more fragmented. Smaller entrepreneurial companies are specializing in research aimed at providing NCEs, or they are becoming experts in managing clinical trials. So the industry may be subject to what Slywotzky (1996) calls *value migration*: the main value in the industry may no longer be produced in the major pharmaceutical companies; instead, they become the 'end producers', turning promising NCEs into final products. To what extent do these trends affect the research environment and aspects of organizational culture? This topic has been debated widely because there are concerns over how to cope, manage, and encourage the organization in order to preserve and support creativity that can lead to innovations (Vrettos, 1999). For example, there is a danger in concentrating on computing power, high-throughput screening, and *in vitro* models, because bioinformatics technology might outpace biology, which risks becoming a 'black box' whose inner workings will be understood by only a few (see Chapter 4 for a more detailed discussion).[10]

The long development cycles in the pharmaceutical industry make it difficult to separate cause and effect. Investments made at the present time will eventually bear fruit in future decades. Consequently, products that have been delivered during the past 15 years, and which made the industry what it is today, were the outcome of scientific innovations

---

[10] *The Economist* (2002) 'The race to computerise biology', 12 December issue.

derived more than 15–25 years ago (Schmid and Smith, 2002b). According to Drews (2003), the closeness to medical biological science and a willingness to submit to the rigour and discipline of good science that used to signify the industry, are now being replaced by a marketing dogma in which R&D is degraded to a tool for generating medicines that qualify as blockbusters. All of these factors have resulted in increased focus on how to manage *organizational creativity* as an organizational capability within the pharmaceutical industry (e.g., Thompson, 2001; Dollery, 1999).

What creativity can do about it

The pharmaceutical industry is investing increasing amounts in R&D and new product development, but have a pay-back problem in terms of diminishing returns on investment in R&D. Creativity is the capacity to provide new solutions and new ideas within the realm of practice. Being able to manage the creativity of the co-workers, pharmaceutical companies may attain a better return on their investments. The notion of creativity is here a 'black box', a series of interrelated practices, cognitive models, laboratory operations, standard procedures, and so forth, whose internal relationships and co-dependencies are not always easily determined. It is therefore complicated to claim that a certain firm need 'more creativity' as if creativity was some standardized activity that could be applied to cases. Instead, creativity is a theoretical construct that can be employed in conversations and discussions to capture those processes that are innate to a successful R&D or new product development product. In other words, creativity per se can promise nothing, but the notion of creativity may be helpful when identifying a number of managerial objectives in the pharmaceutical industry safeguarding a continuous production of new blockbuster drugs. As a consequence, to investigate into the notion of creativity is not only a matter of academic concern but is also of practical importance.

The theoretical perspective: creativity and its competing concepts

In comparison to other domains of organization theory which deal with the use of intellectual resources and know-how, the literature on organizational creativity remain relatively marginal. Before moving into the analysis of the corpus of literature on creativity in organizations (see Chapter 2), some of the competing analytical frameworks will be examined. The point worth considering here is that what the literature on creativity in organizations and management of creativity is trying to capture is in many cases addressed elsewhere – albeit under different labels. In the following, a brief overview of the literature on knowledge management, innovation and R&D literature, organization learning, and entrepreneurship literature will be offered.

Knowledge management

Knowledge management has emerged as a significant new discourse within the wide and diverse field of management studies. One may identify at least two different traditions within the knowledge management literature. First, the notion of the knowledge-based view (KBV) has been suggested as an outgrowth from the resource-based view (RBV) of strategy. In the RBV literature, one of the most significant contributions to strategic management theory in the 1990s, organizations and firms are regarded as bundles of heterogeneous resources. Rather than explaining relative performance of the focal firm in terms of its positioning vis-à-vis its competitors in the market, RBV theory investigates the internal resources of the firm. The knowledge-based view is then emphasizing the intellectual resources of firms as being the most significant single explanatory factor (Spender, 1996; Grant, 1996). The second tradition is the more sociological and constructivist literature that explores the production and use of knowledge as a socially embedded and collective undertaking (Alvesson, 2001; Gherardi, 2000; Tsoukas, 1996). In this view, knowledge is not just some resource or tool that can be applied to cases, but is rather the effect of a shared theoretical and epistemological perspective on what is a legitimate form of knowledge. For instance, a stock trader's know-how of his or her industry is of necessity dependent on the other stock traders' know-how and activities. No knowledge is then knowledge per se, but is always already located within specific communities of practice (e.g., the stock traders' community). These two orientations, the strategic and the sociological view, tend to cluster around two end positions of the continuum; on the one side some knowledge management theory conceives of knowledge as being a stock of skills, know-how, experiences, and expertise that constitute the firm's intellectual capital (e.g., Bontis, Crossan and Hulland, 2002; Von Krogh, Ichijo and Nonaka, 2000; Szulanski, 1996); on the other side of the continuum, knowledge is regarded as what is emerging in practice, grounded in a heterogeneous body of social, emotional, embodied, and intellectual resources (Chia, 2003; Tsoukas and Vladimirou, 2001). In the latter perspective knowledge is what is inherently fluid, provisional, and becoming. As a consequence, what is treated as knowledge is never uncontested, but is rather what can be referred to as a social accomplishment (Orlikowski, 2002). In the field of knowledge management, there is continuous debate between the two (ideal typical) positions.

The knowledge management literature essentially addresses the same issues as the creativity literature. The notion of 'knowledge creation' is perhaps the most closely associated construct that explicitly aims to understand how new ideas and thought are established in organizations. While the knowledge management literature has been in vogue since the end of the 1990s, the creativity literature remains relatively marginal in organization theory.

Innovation, R&D and new product development literature

Research on innovation, new product development and the management of R&D in organization has become one of the largest empirical domains in organization theory and management studies. Since the publication of Burns and Stalker's (1961) classic study of the Scottish knitwear industry, innovation has been a part of the management studies agenda. There is a plethora of theoretical perspectives in this body of literature ranging from a narrow micro level of individuals to the analysis of industrial clusters and regions. Some contributions have bearings on adjacent theoretical fields such as knowledge management (e.g., Subramaniam and Venkatraman, 2001; Von Hippel, 1998) or human resource management (Bunce and West, 1996). Among the various perspectives pursued, the impact of local cultures on innovation (Jassawalla and Sashittal, 2002), the ability to share knowledge within new product development teams (Leonard-Barton, 1995), as well as between business units and firms (Nobel and Birkinshaw, 1998; Powell and Smith-Doerr, 1996), and managers' cognition and conceptualization of innovation (Salaman and Storey, 2002) have been investigated. Different contributors suggest that innovation is an outcome from a superior ability to combine existing resources (Galunic and Rodan, 1998) or to establish an culture that supports experimenting and innovative thinking (Thomke, 2001; Kamoche and Pina e Cunha, 2001), or the ability to manage expert knowledge (Blackler, Crump and McDonald, 1999). Some researchers have claimed that innovation is of necessity at least partially chaotic (Cheng and van de Ven, 1996) while others stress that innovation may very well take place in mature and less fluid organizational forms (Dougherty and Hardy, 1996). Others have stressed that innovation is increasingly the outcome of collaborative efforts between firms, often in the form of network organizations (Jones, 2000; Hage and Hollingsworth, 2000). Even though contributors to innovation such as O'Shea (2002) makes references to process philosophers such as Henri Bergson who remains one of the major thinkers of creativity, there is surprisingly little in the innovation, new product development and R&D literature about creativity *qua* creation. The processes preceding any innovation are, in most cases, not addressed as being dependent upon creativity but are explored through other theoretical frameworks. Nevertheless, there are important results and findings from the innovation literature that are of importance for the research on organizational creativity. Therefore, the literature on innovation, new product development and R&D management need to inform the field of organizational creativity.

Organization learning literature

The notions of organization learning and the learning organization have their roots in the classic organization decision-making literature (March and Simon, 1958) but since the early 1990s the literature on learning in organizations has grown massive. Several literature reviews have been published (e.g., Easterby-Smith, 1997; Dodgson, 1993; Huber, 1991) seeking to outline the basic categories of the field. Learning organization theory draws on a broad array of disciplines such as psychology (Argyris and Schön, 1993), organization theory, sociology (Gherardi and Nicolini 2001), and anthropology (Czarniawska, 2001). There is a significant theoretical diversity in the organization learning literature. For instance, organization learning is conceived of in terms of systems theories (Cohen and Levinthal, 1990; Nevil, DiBella and Gould, 1995), as a subset of the knowledge management literature (Garrick and Clegg, 2000; Selen, 2000), or strategic management theory (Appelbaum and Gellagher, 2000; Winter, 2000), or as being closely entangled with the firm's innovative capacities (Gunther McGrath, 2001). Other contributors emphasize that learning is always embedded in local organizational cultures (Cook and Yanow, 1993; Fiol and Lyles, 1985) as well as norms and cognition that influence and determine activities (Huzzard and Östergren, 2002; Bain, 1998). A more critical stream of analyses examine organizational learning as a highly contested concept embedded in power relations within organizational (Contu and Willmott, 2003; Driver, 2002; Pritchard, 2000) or take a critical view of the construct of organization learning as such (Contu, Grey and Örtenblad, 2003; Levinthal and March, 1993; March, 1991). Several researchers have argued that organization learning must always be examined in terms of being based in practical undertakings and standard operating procedures (Gherardi, 2000; Bryans and Smith, 2000; Tranfield, 2000) and therefore there are some examples of papers suggesting various practices, methods, and tools than support organizational learning (Friedman, Lipshitz and Overmeer, 2001; Pawlowsky, Forslin and Reinhardt, 2001).

Empirical studies of organization learning comprise studies of process industry (Woicehyn, 2000), manufacturing industry (Huzzard, 2000), healthcare organizations (Lipshitz and Popper, 2000), the construction industry (Vakola and Rezgui, 2000), and skilled artisan work (Cook and Yanow, 1993). Other empirical studies have examined learning in specific organizational forms such as virtual organizations (Hedberg and Holmquist, 2001), virtual teams (Sole and Edmondson, 2002) or communities of practice (Wenger, 2000; Tosey, 1999). On the basis of this literature, one can formulate two propositions. For the first, organization learning is always processual, that is, it is emerging and is an outcome from the flow of activities that make up everyday work. Here learning is an effect of the series of interrelated practices and operations that are undertaken daily. Gherardi and Nicolini (2001: 35) talks about this processual view as *learning-in-organizing*: 'Organization learning is learning-in-organizing because working learning, and organizing are not distinct activities within a practice. The concept of participation, therefore, gives access to the study of organizational learning that takes place in action and through action.' Organizing is itself a process of ordering, of bringing together different activities and resources, and learning is an integral component of that process. Secondly, organization learning is relational – that is, learning is what appears and emerges in relationships between human beings, between peers, within communities of practices, or within professional groups. Weick and Westley (1999: 196) write:

> Learning is embedded in relationships or relating. By this we mean that learning is not an inherent property of an individual or of an organization, but rather resides in the quality and the nature of relationship between levels of consciousness within the individual, and between the organization and the environment. Thus learning at the individual level (interpersonal) and at the organizational level (interpersonal or interorganizational) evolves through a continuous process of mutual adjustment.

Learning is here thus not simply located within individuals' minds or bodies, but is distributed across different actors. This brings us to the third aspect of organization learning which emphasizes the importance of artefacts in process of learning. Since learning is relational it would be problematic to argue that learning is solely a matter of relations between humans. Instead, learning is dependent on the active use of non-human resources such as artefacts. Professional and skilled work is inextricably entangled with the use of different tools, machinery and other artefacts – for instance, computers or hammers. Processes of learning are, in most cases, closely associated with the use of these artefacts and therefore they cannot be excluded from an analysis of organization learning. Cook and Yanow (1993), studying the production of state-of-the-art musical instruments, flutes, in three Boston flute companies, argue that processes of learning are always related to the materiality of the flutes. As a consequence, all learning emerged when embodied, tactile and audial capabilities of the flute-builders interacted with the physical artefacts, that is, the flutes, and the conceptual framework mastered by the flute-builders. Some flutes were not really acceptable for some co-workers because they 'did not feel right' and then the flutes had to be corrected. Cook and Yanow (1993) here define organization learning as: '[t]he acquiring, sustaining, or changing of intersubjective meanings through the artifactual vehicles of their expression and transmission and the collective actions of the group' (Cook and Yanow, 1993: 384; original in italics).

In summary, the organization learning literature offers a framework which allows us to understand how organizations adapt to changes in the external environment and how they develop practices for sharing know-how within the organization. The organization learning literature is therefore of relevance for the understanding of organizational creativity.

Entrepreneurship literature

One final empirical domain of investigation that shares interests with the creativity literature is research on entrepreneurship. The notion of the entrepreneur was introduced in Austrian economic theory and was popularized by the Harvard professor Joseph Schumpeter who argued that the entrepreneur is the key economic actor in terms of inventing new ways of doing business. In Schumpeter's parlance, the entrepreneur – and his or her ability to conceive of new business opportunities – enabled creative destruction of the existing economic structure; the entrepreneur may alter the corporate landscape through changing the rules of the game. In today's western economies, struggling with unemployment and diminishing rates of growth in the economy, the entrepreneur is becoming an increasingly praised figure, embodying all the momentum of capitalist dynamics, and who is bringing the hope of new job opportunities and new goods and services. One may therefore argue that the entrepreneur is becoming a somewhat mythological species of the capitalist landscape (Ogbor, 2000; Mitchell, 1996). From the outset, the entrepreneur is regarded a creative person (Hjort, 2003), somebody who can spot a business opportunity before anyone else. Entrepreneurship research is often taking its point of departure in strategic management research (see, e.g., Hitt, Ireland, Camp and Sexton, 2001), but is becoming very much a discipline of its own. Even though the notion of entrepreneurship is treated as an activity that takes place on green fields, previously unaffected by economic activities, there are a number of studies that investigate entrepreneurship in large organizations (e.g., Ahuja and Lampert, 2001). The term *intrapreneur* has, in fact, been coined to denote this particular form of entrepreneur, operating within stable structures.

Concluding remarks

There are a large number of alternative theoretical perspectives that seek to understand – or even explain – why some organizations and firms are more successful than others in exploiting their intellectual resources. The knowledge management literature, the writings on organizational learning, innovation and new product development literature and the discourse on entrepreneurship are four examples of theoretical domains within organization theory and management studies that explore processes in organizations and firms that may be appropriately labelled as being based on creativity. This corpus of literature is complementary to the writings on creativity, but it also offers four examples of vivid theoretical debates from which the management of creativity can learn. In comparison, the literature on creativity is relatively underdeveloped; it is less frequently referenced in academic management journals and books, it provides less systematic research, has fewer standing conferences and research communities, and is less visible to the mainstream of management studies. This is unacceptable because the notion of creativity is a handy construct when seeking to understand how science-based innovation such as in the pharmaceutical industry is taking place. If being given a more elaborated theoretical framework, the notion of creativity is promising to offer good opportunities for thinking of innovation, knowledge, and management in different terms. Making the construct of creativity a source of investigation is therefore one of the key priority in this book.

Finally, the innovation literature is extensive but lacks understanding and connection to its precursor: organizational creativity. Furthermore, organizational creativity is an emerging field of research, but is lacking in empirical studies grounded in an organizational context. Besides, as Mumford (2000) argues, the need for innovation has sparked a new cottage industry proffering advice and 'how-to' manuals on management practices that should be applied to encourage creativity. Unfortunately, as pointed out by Montouri (1992), few of these efforts consider available research that examines the nature of creativity. As such, this book is a contribution to the organizational creativity literature and a link to innovation research.

Outline of the book

This book is organized into two distinct parts. The first part focuses primarily on an analysis of the construct of creativity in terms of its ontological and epistemological basis, its relationship to organization theory and management studies, and the methodological concerns associated with the study of creative work. The second part suggests a number of empirical illustrations, based on a combination of case studies and quantitative methods such as survey methodology, of what challenges the management of creativity is facing in the day-to-day practice. To make the point more clear, the first part aims to establish creativity as a legitimate scientific construct through a systematic critique of its implicit assumptions, while the second part addresses a number of practical aspects of creativity. These two perspectives are, albeit being separated into the two parts of the book, interrelated and mutually dependent throughout the book: Without a proper and robust theoretical framework, an analysis of practices is prevented; without a firm empirical basis, the theoretical constructs remain hollow signifiers detached from any managerial practice in organizations and companies. In all research activities, there is co-dependent relationship between what Hacking (1983) calls *representing* ('theory') and *intervening* ('empirical research'). As a consequence, the reader should not treat the two parts as being wholly separated units but should rather conceive of them as being complementary. However, it is possible to read the two parts of the individual chapters on their own. In the final chapter, we discuss in more practical terms how organizational creativity can be managed in organizations and what implications factors such as technology, leadership and cognition have for the exploitation of organizational creativity.

Conclusions

This introductory chapter has argued that the literature on creativity in organizations and the management of creativity offer opportunities for further analysis and discussion. In comparison with other theoretical frameworks, the notion of creativity occupies a rather modest position within management literature. In addition, managers may have a practical use of a notion of creativity that is more theoretically developed and that denotes a relatively stable set of activities and propositions. As a consequence, an analysis of the notion of creativity is therefore beneficial both for academic management researchers and for practitioners.

Part I
Critique

2
What is Organizational Creativity?

Introduction

This chapter will consider the literature on organizational creativity. This corpus of literature is rather heterogeneous, comprising contributions from such diverse disciplines as psychology, management studies, sociology, and even the humanities (literature theory and history). The scope of the literature is therefore rather broad, ranging from, on the one hand, laboratory research on decision making under controlled conditions and, on the other, biographically oriented essays on the work of individual artists. The literature on creativity is then not unified and integrated, but may instead be regarded as a loosely grouped series of statements and propositions on what creativity is, what its consequences are, and how it can be employed purposefully. In addition, the idea of creativity is very much playing a role in the public imaginary as some superhuman capacity to conceive of extraordinary things which certain individuals are endowed with. As a consequence, the theoretical analysis of creativity *qua* theoretical concept needs to address this everyday view of creativity as something that occurs disruptively in unpredictable occasions among extraordinarily talented individuals. In fact, this mythological image of creativity is counterproductive to the endeavour of managing creativity as an organization resource similar to knowledge or technology because it postulates that creativity is highly unpredictable.

Creativity research: a literature review

Creativity is one of the most intriguing and elusive topics to be associated with human performance (Ford and Gioia, 1995). Creativity is also one of those words that seem to be everywhere. Creativity has many meanings, which often are not made explicit enough to avoid confusion and thus impede communication (Feldman *et al.*, 1994). The terms *creativity* and *creative* also have highly positive connotations. The study of creativity has always been tinged or tainted with associations with mystical aspects (Sternberg and Lubart, 1999). In addition, much research has in some way been influenced by the *romance of creativity*, a factor which may oversimplify explanations of events and attribute great creative achievements to single individuals (Ford and Gioia, 1995; Isaksen, 1987). According to Ford (1995b), creativity is a subjective judgement of novelty and value often studied in non-working areas. Creativity is a multifaceted concept that is manifested in different ways in different domains and it acquires different meanings for different organizations (e.g., Rickards, 1991; Runco and Pritzer, 1999). This pluralism in perspective makes creativity difficult to define and problematic to assess. During more than 50 years of creativity research, many researchers and consultants have treated the concept as an individual trait and have paid little attention to the organizational context or professional concerns (Ford, 1995a), thus underestimating its social and organizational components (Csikszentmihalyi and Sawyer, 1995).

Research on creativity has struggled in seeking generalizability and validity in defining creativity (Plucker and Runco, 1998). According to Rickard and De Cock (1999), one problem is that many definitions become trapped within closed systems of completely self-referential concepts. For example, a creative product is viewed as creative since it is the outcome of creative processes; a creative person is recognized by his or her creative products and the creative processes he or she experiences. Because creativity has been viewed as an all-encompassing construct, studying any one part in isolation then decreases the validity of the construct. However, as Schoenfeldt and Jansen (1997: 74) write: 'By considering both creativity and innovation research and by adopting an interactionist approach, the linkage between process, product, person and situation are retained and new methods for studying creativity may be found.' According to Mayer (1999), there is a consensus among researchers about *two* defining characteristics of creativity: *originality* and *usefulness*. In this sense, Sternberg and Lubart's (1999: 3) definition is an appropriate example: they define creativity as the ability to produce work that is both novel (i.e., original, unexpected) and appropriate (i.e., useful, adaptive concerning task constraints).

In much of the research and management literature, the terms *creative* and *innovative* often tend to overlap. Some researchers assert that the distinction between innovation and creativity may in reality be more of a case of emphasis than one of substance (Scott and Bruce, 1994; West and Farr, 1990). One distinction that can be made between them is to consider creativity as the *generation of ideas* for new, improved ways of doing things and innovation as the *implementation of those ideas* in practice (West, 1999). In general, creativity is perceived in highly individual terms and as something that only expresses itself fully in non-work-related areas. It is seen as a process that can be facilitated by ways of working and thinking (Williamson, 2001). Therefore, innovation is often associated and sometimes confused with the related concept of creativity (Ford, 1996). According to Magyari-Beck (1999), creative products are at the level of organizational innovations. Woodman *et al.* (1993) refer to creativity as a subset of innovation and Amabile (1988) views creativity as a necessary precursor to innovation, thus defining creativity as the production of novel and useful ideas by an individual or small group of individuals working together. According to Cropley (1999), creativity can be defined as the production of relevant and effective novelty. Innovation can then be seen more as the process of the implementation of, and decisions about, creative ideas (West and Rickards, 1999; Shani and Lau, 2000). Or, as Mumford *et al.* (2002: 708) conclude, 'with the generation of new ideas, the idea development and implementation activities that characterize innovation become possible'.

Most accounts of innovation describe a novel idea that appears at the start and gets introduced into practices as an innovative product that can take a variety of forms. Moreover, innovation is predominantly seen as technological process operating in a closed system which is presumed to evolve step by step in linear and causally connected stages (e.g., Cooper, 1992). This view is well exemplified by Toffler's (1972) definition of technological innovation which consists of three stages linked together into a self-reinforced cycle (first, there is the creative, feasible idea, followed by its practical application, and third, diffusion through society). This view makes the innovation process connected with creativity but 'at the front end'. This 'front-loading' of creativity means that the whole tricky question of the discovery processes has been eliminated, so the subsequent stages can be presented as rational and logic sequences of activities. Creativity, if addressed at all, is isolated and controlled. According to Rickards (1999: 47), important consequences flow from isolating creativity in this way: 'It inexorably leads to so-called stage models of innovation in which uncertainties of discovery are decoupled from models of innovation in which uncertainties of discoveries are decoupled from the later stages of the system.'

Four perspectives on creativity

The beginning of the field of creativity research is usually marked by J.P. Guilford's presidential address to the American Psychological Association in 1950 (Guilford, 1950).[11] Guilford's basic approach to creativity was to isolate various traits of intellect and personality that 'creative' individuals might possess in greater quantities than others and to demonstrate that creativity is a dimension separate from intelligence (Guilford, 1950). Thus, Guilford laid a foundation for modern research on the creative personality (Feldman *et al.*, 1994). Since then, the research on creativity has emerged from many academic disciplines, including psychology, organizational behaviour, education, history, sociology, and various management disciplines. The vast research literature on creativity is often either person-centred or focused primarily on specific aspects of creativity (Ford, 1995a). These specific aspects have been thoroughly researched and have included primarily four distinct aspects of understanding creative acts. These specific areas of creativity research include the creative processes, the creative person (i.e., personality and behavioural correlates), creative products (i.e., characteristics of creative products), and the creative place (i.e., attributes of creative supporting environments).

The *creative process* research stream was person-centred, aiming to quantify the creative process primarily through use of divergent thinking batteries and cognitive variables, such as thinking styles, skills, and problem-solving techniques (Plucker and Renzulli, 1999). This line of research aims to develop models and tests in order to provide a comprehensive theoretical framework of how creative thinking and problem-solving can be understood. The SOI (Structure Of the Intellect) model developed by Guilford (1967) and the TTCT (Torrance test of creative thinking) developed by Wallach and Kogan (1965) are classical examples of models for evaluating the creative process by using different constructs (e.g., flexibility, fluency, redefinition, and sensitivity of creative thinking) and different psychological operations (e.g., memory, cognition, divergent thinking, and convergent thinking, or evaluation) (Michael, 1999).

---

[11] According to Albert and Runco (1999: 17), one of the most widely cited statements from Guilford's article states that, of 121,000 cited articles in *Psychological Abstracts* from the late 1920s to 1950, only 186 dealt with creativity.

The *creative person* research was focused on measuring facets of creativity associated with creative people, including personal properties, traits and behaviour in terms of the ability to generate new ideas (e.g., self confidence, tolerance of ambiguity, energy, desire for independence, playfulness, and domain knowledge) (e.g., Amabile, 1984; Eysenck, 1996 and Gardner, 1994). Research methods and instruments were mainly designed to study highly creative individuals and to determine their common personality characteristics (Plucker and Renzulli, 1999). Studies were primarily idiographic or nomothetic research. The idiographic research focuses sharply on individual case studies with their emphases and wrinkles. Typical examples are studies of highly creative individuals such as Charles Darwin or Jean Piaget (Gruber, 1981). In nomothetic research, the focus instead falls on a search for general laws and attempts to overlook individual idiosyncrasies, searching for patterns that appear to apply to all or to the majority of cases (Gardner, 1994). A classical example is Simonton's (1984) historimetric study of highly prominent and creative people from the past, including politicians, philosophers, scientists, and writers.

The third research domain of creativity, *the creative product*, deals, for example, with aspects of the evaluation of what defines creative output (e.g., originality, relevance, usefulness, complexity, and how pleasing the output is). MacKinnon (1978: 187) argued: 'The starting point, indeed the bedrock of all studies of creativity, is an analysis of creative products, a determination of what makes them different from more mundane products.' The aim is here to develop methodologies where creative products are rated by external judges using sets of traits, such as originality, appropriateness of resources, and audience. (Plucker & Renzulli, 1999). But the instrument has built-in methodological problems, such as validity and reliability and the general criterion problem of creativity. A classical example is Amabile's (1982) consensual assessment technique (CAT). The CAT technique partially overcomes these problems by using an amorphous definition of creativity: 'A product or response is creative to the extent that appropriate observers independently agree it is creative' (Amabile, 1982: 1001). Thus criterion problems are avoided, individual differences are eliminated, and environmental influences can be better examined (Plucker and Renzulli, 1999).

The last major research stream, *the creative environment*, focuses on the study of the context in which creativity occurs; it aims to investigate different environmental variables related to creative productivity. By studying contextual aspects of creativity, the focus also changes from the dominant individual-centered psychometric perspective toward an organizational and managerial context. Amabile's work (1988, 1997) constitutes one example of this line of research that involves social, ecological, and contextual factors. Amabile proposes a model that consists of five environmental components which affect creativity in organizations: *encouragement of creativity* (information and support for new ideas must be communicated openly between all different levels in the organization); *autonomy* (individual freedom and control must be an integral part of day-to-day work); *resources* (basic materials and information for the work must be available); *pressures* (positive challenges must be imposed and negative perceptions of workloads should be avoided); and *organizational impediments to creativity* (influences of conservatism and internal strife must be reduced). Another example is referred to as the *creative climate* of an organization (Ekvall, 1996, 1997). This includes an organization's leadership styles, visions, objectives, goals, strategies, resources, personnel policies, beliefs, values, structures, and systems. All these factors are crucial for how people view the climate in which they work. As an example, there is an important link between the creative climate and innovation (Ekvall, 1987, 1995). By diagnosing the creative climate, one may distinguish between innovative, average, and stagnated organizations—based on product performance and the success of the organization as a whole. Other studies also offered empirical support for a relationship between perceived climate and innovation (Abbey and Dickson, 1983; Paolillo and Brown, 1978; Siegel and Kaemmerer, 1978).

What is organizational creativity?

The traditional strong individual focus and the *romance of creativity* (Ford & Gioa, 1995: 3) are not well suited to understanding and promoting creativity in organizations. According to Ford (1995a), the dominant approach when explaining creativity is to think of personal qualities, such as intelligence or divergent thinking skills and other heroic qualities and ignoring the context within which creative products emerge. Another aspect of 'traditional' creativity research is that the distinct foci of creativity do not give a useful understanding of how creativity works in an organizational context. Furthermore, much of what passes for common wisdom about managing creativity is drawn from research from non-organizational domains (e.g., fine and performing arts, education, the history of science, and child development) (Ford, 1995a). Recently, the issue of creativity has been viewed as an important organizational resource (Ford and Gioa, 2000; Kazanjian *et al.*, 2000; Mumford *et al.*, 2002; Williamson, 2001). While there is a strong tradition of studying creativity as a personal characteristic of individuals, conditions that promote creative performance in organizations remain largely unknown (Oldham and Cummings, 1996). In recent research, steps were taken to understand creativity in an organizational context, using concepts such as *organizational creativity* (e.g., Basadur, 1997; Clitheroe *et al.*, 1998; Ford, 1996; Woodman *et al.*, 1993). But there is still a paucity of research material on the subject (Ford and Gioia, 1995) which indicate a need for research that will suggest how organizations should be designed to facilitate the flow of ideas that lead to innovation (Jacob, 1998). According to Csikszentmihalyi (1994), one must widen the view of what the process is, moving from an exclusive focus on the individual to a systemic perspective that includes the social and cultural context in which the 'creative' person operates. According to Ford (1995a), creativity is not an inherent quality of a person, process, product or place but is rather a domain-specific social construction that is legitimized by judges who serve as gatekeepers to a particular domain. Thus, an important step in understanding creativity in an organizational context is to take a more holistic approach and use the concept of *organizational creativity*. A somewhat similar concept is *corporate creativity* (Robinson and Stern, 1997) which specifies creativity within a company context. According to Robinson and Stern (1997: 17) corporate creativity can be defined as when its employees do something new and potentially useful without being shown or taught.

So what is organizational creativity? One useful definition is offered by Woodman *et al.* (1993: 293), according to whom it is 'the creation of a valuable, useful new product, service, idea, procedure, or process by individuals working together in a complex social system'. Organizational creativity emphasizes social and group creative processes (Csikszentmihalyi and Sawyer, 1995). But organizational creativity can also refer to the extent to which the organization has instituted formal approaches and tools and provided resources to encourage meaningful novel behaviour in the organization (Bharadwaj and Menon, 2000). Thus, organizational creativity can be seen as a phenomenon that is structurally embedded in the organization rather than being some innate quality of a few extraordinary individuals, as Jacob (1998) insists, emphasizing that organizational creativity is something more than a collection of creative individuals. To be able to acknowledge the context-specific aspects of creativity in organizations, it must be articulated in terms of the organization's mission and cannot only represent novel acts. It must produce value relative to an organization's mission and market. This means that creativity in organizations is valuable only if it is implemented in such a way that it is adapted to the organization's culture, values, and processes (Gioia, 1995) or if it turns company values and processes upside down. Organizational creativity can also be seen as a subset of two broader domains: organizational change and innovation (Kilbourne and Woodman, 1999).

When seen from a practitioner's perspective, these aspects fall into a definition of organizational creativity that comprises acts of envisioning, demonstrating, and applying cost-effective methods for the purpose of eliminating technological problems and providing significant, profitable technology-based opportunities in target areas of business activity (Jones, 1995). So the concept of organizational creativity opens up new linkages to other organizational aspects – for example, organizational learning (Ford, 2002; Koh, 2000), information and knowledge management (Styhre and Sundgren, 2003b), leadership (Mumford *et al.*, 2002), and entrepreneurship (Hjort, 2003).

To summarize, creativity is believed to be important by a majority – if not all – of the organizational practitioners and managers and consultants. Despite this, creativity still stands outside of management studies and is not taken seriously (Rickards, 1999). An overview of the literature on creativity demonstrates that the concept is still fragmented, compartmentalized and suffers from the individualist view of the psychological research heritage. In addition, research and literature on creativity is still lacking an integrated view of the complex landscape of creativity in organizations (Ford, 1995). Moreover, the vast literature on innovation either chooses to ignore the concept or simply encapsulates creativity into controlled stages (as rational and logic sequences of activities) in which much of the innovation models predicts. In short, empirical studies of creativity in organization, or organizational creativity are scarce.[12]

Creativity as kitsch and management fad

Since the early 1970s research on creativity has been dominated by psychological interpretations. Perhaps as a consequence, traditional approaches to the study of creativity have been focused on the main,

---

[12] As one example, a search in the *Social Science Citation Index* (from 1975 to November 2003) gave 10011 matches for the topic 'creativity', 29423 matches for the topic 'innovation', and 21 matches for the topic 'organizational creativity'. Of these 21 articles, less than six were empirically based.

and often main contributor to creativity (Ford and Gioia, 1995). In addition, the literature on innovation is prolific, but lacks an appreciation of, and connection to, its precursor, organizational creativity, which is an emerging field of research but is largely lacking in empirical studies grounded in an organizational context. Besides, as Mumford (2000) argues, the need for innovation has sparked a new cottage industry proffering advice and 'how-to' manuals on management practices that should be applied in an attempt to encourage creativity. This is also reflected in the mainstream management literature on creativity, which to a large extent presents a simplified picture of the management aspects of creativity in organizations.

On the other hand, as Jung (2001) pointed out, leadership – at least traditionally – has not been held to be a particularly significant influence on creativity and innovation. One reason for this tendency to discount leader influences may be found in the romantic conception of the creative act – a conception in which ideas and innovation are attributed to the heroic efforts of the individual (Amabile *et al.*, 2004). More specifically, one can argue that the professionalism, expertise, and autonomy that seem to characterize creative people act to neutralize – or substitute for – leadership (Mumford, Scott, and Gaddis, 2002).

In the following, we use the notion of *kitsch* as a theoretical construct, aiming to capture two specific qualities: (i) the problem of transporting one tradition of thinking into a second domain; and (ii) the emphasis on wishful thinking on what the world could possible be. These two qualities are representative of an ideological and aesthetic position that has been labelled kitsch in the literature on aesthetic works (literature, visual arts, marketing material, etc.). Before moving on to the corpus of organizational creativity literature, we will engage with a more detailed account of the notion of kitsch.

The *locus classicus* for a more detailed analysis of kitsch is Hermann Broch's essay (1933/1868). Commenting on the text, the Czech author Milan Kundera (1988: 135), himself an explorer of kitsch (Kostera, 1997), writes:

> For Broch, kitsch is historically bound to the sentimental romanticism of the nineteenth century. Because in Germany and central Europe the nineteenth century was far more romantic (and far less realistic) than elsewhere, it was there that kitsch flowered to excess, it is there that the word 'kitsch' was born, there that it is still in common use.

For Umberto Eco (1989: 198), kitsch is to be regarded as a substitute for real art: 'As an easily digestible substitute for art, Kitsch is the ideal food for a lazy audience that wants to have access to beauty and enjoy it without having to make much of an effort.' The same kind of derogatory remarks are provided by Walter Benjamin (1999: 395): 'Kitsch... is nothing more than art with a 100 percent, absolute and instantaneous availability for consumption. Precisely within the consecrated forms of expression, therefore, kitsch and art stand irreconcilably opposed.' Kitsch is then what is a simulation of art but which fails to achieve the same degree of *authenticity* with the real. Kitsch is then what is running parallel to regular art, seeking to become art but always succumbing to commercial forces and commonsense thinking. In our view, there are two characteristics of kitsch that are of particular interest. For the first, kitsch is what emerges when one aesthetic element is transferred into a new domain; '[kitsch] occurs each time a single element or a whole work of art is 'transferred' from its real status and used for a different purpose from the one for which it was created' (Dorfles, 1968: 17). This aspect of kitsch is shared by Eco (1989: 201): '[I] would like to define Kitsch in structural terms, as a styleme that has been abstracted from its original context and necessity as the original's, while the result is proposed as a freshly created work capable of stimulating new experiences.' In this view, kitsch implies a movement of aesthetic elements into places in which it was not originally located. Examples abound: Mona Lisa's face on a package of cheese (Dorfles, 1968: 19); Venetian squares in the Nevada desert; garden gnomes – symbols of pre-modern folk beliefs – in modern suburban gardens. This is the most generic definition of kitsch – as what is *dislocated*.

The other aspect of kitsch is the ideology of wishful thinking and rosy images of the real, detached from everyday life experiences. Hermann Broch (1933/1968: 62) writes: 'Kitsch is certainly not "bad art"; it forms its own closed system, which is lodged like a foreign body in the overall system of art, or which, if you prefer, appears alongside it'. He continues:

> The essence of kitsch is the confusion of the ethical category with the aesthetic category; a 'beautiful' work, not a 'good' one, is the aim; the important thing is an effect of beauty. Despite its often naturalistic character, despite its frequent use of realist terminology, the kitsch novel depicts the world not as 'it really is' but 'as people want it to be' or 'as people fear it is'. (Broch, 1933/ 1968: 71)

For Broch, kitsch is what fails to address the state of things but rather indulges in free-ranging fantasies of how 'people want things to be'. Kundera (1988: 163) develops this line of thought:

> The word 'kitsch' describes the attitude of those who want to please the greatest number, at any cost. To please, one must confirm what everyone wants to hear, put oneself in the service of received ideas. Kitsch is the translation of the stupidity of received ideas into the very language of beauty and feeling. It moves us to tears of compassion for ourselves, for the banality of what we think and feel.

In an attempt to further discuss the notion of kitsch within organization theory, Stephen Linstead speaks of this tendency to please the multitude the 'narcissistic properties of kitsch': 'The narcissistic properties of kitsch, and the tendency of the familiar to follow a trajectory of deepening approval from the aesthetic (it is comfortable, pleasing) to the moral (it is approved, advocated, required, the natural way of things) underpin a cosmology which positions humanity at the centre of creation' (Linstead, 2002: 664). Linstead continues by pointing beyond the aesthetic implications of kitsch, suggesting it may be a fruitful analytical category for examining social practices and social relations:

> [N]ot only is kitsch a means of achieving cheap artistic effects, it is also a means of achieving cheap social and political effects. Rather than simply selling aesthetic form, kitsch sells ideas and feelings, and the 'bag of tricks' of art becomes available for any purpose. (Linstead, 2002: 666)

Linstead writes: '[I]t is not the identification of kitsch as an aesthetic style in organizing which is significant, but the recognition of kitsch as an ontology of being which effectively masks the experience of being – interposing itself as a comforting buffer between ourselves and the "real", and often being taken for it' (Linstead, 2000: 657). To Linstead (2000), organizational kitsch is something 'that prettifies the problematic, makes the disturbing reassuring, and establishes an easy (and illusionary) unity of the individual and the world'. Linstead argues that the management bestseller *In search of excellence* by Tom Peters and Robert Waterman (1982) is a publication that draws on this particular ethic of organizational kitsch.

In our view, the notion of kitsch apprehends two facets of the organizational creativity literature. First, it points at the swift and often momentary movements between idiosyncratic domains and areas of management practice, using the same kind of management tools and techniques to capture the supposedly innate creativity of the co-workers or production systems. Similar to the concrete replicas of Michelangelo's *David* on sale in the western world for the benefit of garden decoration (Dorfles, 1968), creativity in specific contexts is treated as being universally applicable in all possible domains and therefore possible to transfer into new settings. Secondly, the very idea of creativity in organizations is based on what Linstead calls 'narcissistic properties of kitsch', that is, the one-dimensional emphasis on creativity as what is capable of dealing with all kinds of social and organizational evils. To be fair, it may not only be the literature on organizational creativity that suffers from this 'prettified view'; similar single-handed belief in one single theoretico-practical construct is demonstrated in the entrepreneurship discourse and the writings on information technology (see, e.g., May, 2002). Nevertheless, the organizational creativity literature demonstrates its belief in moving practices between domains and neglect or exclude the political and social costs of creativity. Consequently, creativity is portrayed as what is a quick fix, fun, easy and liberating rather than being the outcome from a tightly knitted system of a political economy of creativity wherein creativity not only has effects but also demands costs and efforts.

One book that serves as a fine illustration of the kitsch view of creativity is William C. Miller's *Flash of Brilliance*, a book aimed at promoting what may be called 'everyday creativity'. In *Flash of Brilliance*, Miller envisages creativity as being tightly connected to the individual's spirit and spirituality. Creativity is therefore what is located within each and every human being, and, Miller suggests, needs to be 'awakened': 'Courageously, and with profound curiosity, we need to awaken the slumbering, creative genius inside ourselves' (Miller, 1999: 5). '*Flash of Brilliance* aims to take you beyond what you thought was your own creative edge', Miller (1999: 6) declares. In terms of kitsch, *Flash of Brilliance* meets the qualifications regarding its first criteria, that is: (i) the problem of transporting one tradition of thinking into a second domain. Written as a typical management guru or 'How to' book, Miller blends a variety of highly heterogeneous resources and brings together all kinds of stories: A series of anecdotes from companies, university courses, and elsewhere, quotes from people like St Augustine (p. 14), Dwight D. Eisenhower (p. 21) and Dalai Lama (p. 13), references to individual experiences and undertakings, and so forth (for a critical review of this genre, see Frank, 1997). Many of these quotes and anecdotes say very little about creativity in organizations, but instead constitute a series of *bon mots* that have been applied to a new domain. For instance, St Augustine is not regarded an authority on organizational creativity, but the status of his work as being foundational for Christian theology is exploited by Miller in what may be called a 'credit by association' stratagem. Analogous to the use of Venetian architecture or Egyptian pyramids in Las Vegas, Miller invokes the great North-African theologist in his text in order to attract interest and build confidence. In addition, *Flash of Brilliance* does not define an operational domain, or industry, or a field, but is speaking of creativity *as such*, thereby largely detached from the idiosyncrasies of different industries that enable or hinder creativity. As a consequence, interesting people like Jonas Salk[13] are quoted without really making sense for the reader because there is little said about in what context such quotes are supposed to make sense. Expressed differently, Miller is sharing with the great utopians the disregard for the conditions of everyday life and the dream of a better world that does not departure from what the world is like, but that is rooted in an imaginary devoid of disturbing mundane practicalities.

The second criteria for kitsch in our account – (ii) the emphasis on wishful thinking on what the world could possible be – is the other domain where Miller excels. What is particularly characteristic of Miller's text is the anthropocentric and evangelical tone he sustains throughout the text. The following quote is typical:

> From the creative contribution of people like yourself, major corporations, grassroots nonprofits, government agencies, and small businesses grow and flourish. We innovate by taking our creative ideas and producing something with them. That's how we renew ourselves and stay healthy to serve our customers, clients, shareholders, other stakeholders, and ourselves. Whether we express our creativity in new products and services, new work processes, or new marketing methods, creativity is the prime source, the taproot, from which solutions spring. (Miller, 1999: 14)

---

[13] Jonas Salk (1914–95), world-renowned US physician and medical researcher who led the development of the first safe and effective vaccine for poliomyelitis (polio) in 1955. In 1963 he became fellow and director of the Institute for Biological Studies in San Diego, California, later called the Salk Institute.

Here Miller subscribes to a consensus view of corporations, a romantic image of companies as sites of social bonding and shared objectives and concerns, wherein creativity is what protects such safe havens from disappearing. The strong focus on individual's potentiality for creativity leads Miller to solipsisms such as in the following claim: 'The ultimate definition of creativity in your life is your personal one' (Miller, 1999: 36). In addition, contrary to the findings of creativity researchers like Sternberg (2003), Miller (1999: 37) argues that there is a low correlation between IQ and creativity. As a consequence, creativity is something for 'everybody' and therefore Miller suggests a certain *modus operandi* for becoming creative; one needs to 'embark on one's creative journey'. Miller writes:

> Let things happen as you would wish them to. Don't inhibit your creative dreaming with thoughts like 'That could never happen', or 'He (or she) would never go for that'. Create the vision, then update it later. That way, you let yourself stretch and grow while still staying 'realistic'. (Miller, 1999: 42)

In this staccato style of writing, filled with list of 'How to' and 'do's and don't's' and other forms of what Osbourne (2003: 508) calls 'technologies of creativity', Miller provides a number of hands-on recommendations and spiritual guidance for individuals wishing to become more creative. It is little wonder, then, given Miller's strong preference for portraying creativity as what carries the potential not only for changing the individual's outlook and for providing new life chances, but also to serve as what is enabling for new a more competitive industry, that Miller ascribe the notion of spirituality a substantial importance in creative thinking. Creativity is here not an outcome from social practices under certain and determinate conditions, but is more of a form of religion and manifestation of spirituality.

Taken as a whole, *Flash of Brilliance* is a book that adheres to the criteria of kitsch that we outlined above. It brings certain resources and materials (e.g., quotes and references to an alarming variety of authorities of all kinds) into a new domain that is alien to those resources. Such resources are exploited in terms of being authoritative, but are at best loosely connected to the topic examined. Secondly, Miller subscribes to an anthropocentric and consensus-based social model wherein every human being has the capacity to become creative. Even though such a vision is appealing and one can have a great deal of sympathy for Miller's belief in human beings, it wholly ignores the material conditions for creativity and how creativity is inextricably entangled with different opinions and debates that may generate new ideas and solutions. Rejecting Miller's firm belief in 'every man's right to be creative' does not imply an elitism but is rather an insistence on regarding creativity in terms that acknowledge its full complexity; creativity does not simply materialize as soon as you follow a number of checklists aimed at making you become 'spirited', but is an outcome from highly specialized activities and a great deal of experience from working in a particular field. In contrast to Miller's view, creativity may be thought of as what demands expertise and experience: We are not all 'creative geniuses'. Most of us have neither the ambition, nor the capacity to become such outstanding contributors – a Leonardo de Vinci, Louis Pasteur or Jonas Salk. Denying this reality is a distinguishing mark of the evangelical management guru writer ignoring the social realities making up everyday life.

Such kitsch management writing is unfortunately of little practical use. For example, improving understanding of how to manage organizational creativity in pharmaceutical R&D has little to do with applying – or relying on – pragmatic creative thinking techniques represented, for example, by de Bono (1985, 1992) or by the plethora of popular management literature (such as self-help books on how to increase one's creativity). The assumption behind this is threefold. First, the concern in these techniques is not with theory, or testing validity, but with practice. The primary goal is to provide toolkits for producing divergent ideas *as such*. Moreover, these activities often take place as conferences outside the organization, placing little or no emphasis on an organizational context of creativity. Secondly, there is an underlying assumption behind these commercialized tools that creativity is an activity that cannot occur under normal working circumstances; one must be trained away from the organization. Robinson and Stern (1997) for example, argue that creativity methods (i.e., brainstorming) may actually limit people's creativity by removing them from their workplace. Thirdly, there is no real empirical evidence that these kinds of techniques really matter in an organizational setting (e.g., Nickerson, 1999; Eysenck, 1996).

Thus, the notion of management and creativity as a management fad drawing on an ideology of kitsch is to a large extent preserved in the main body of this management literature in a rose-tinted and oversimplistic way. So, these commercial creativity techniques and most of the popular creativity literature (e.g. Goodman, 1995; Henry, 2001) offer no added value for an organization that seeks to promote and understand organizational creativity. They merely preserve the ambiguity and romance, designating creativity as a mystical process that only involves the chosen few. However, the management role is highly important for several reasons (e.g., they decide what is creative or not, and if an idea is promising or not to develop further) and is inextricably intertwined with the organizational context. A new direction for management literature on how to manage creativity would be to take a more pluralistic view on how different organizational context play and also considering the more problematic issues of organizational creativity.

Conclusions

This chapter has reviewed the literature on creativity in organizations. This literature includes at least four different perspectives on what creativity is and how it should be studied. This diversity leads to a failure to establish a coherent and unified view of what creativity is and how it can be managed. This does not imply that the pluralist view is mistaken, but it does suggest that it is problematic to speak of creativity as one single resource. Instead, creativity may be regarded as what is appearing in different forms and in different settings. This makes creativity a contingent and situational concept. In addition, some parts of the literature are drawing on certain beliefs and ideologies that we, in this context, have been speaking of as kitsch, a form of overrating the individual's role and function in creative activities and the denial of social costs and efforts involved in creative work. Taken together, the literature on creativity is fragmented and includes a series of alternative perspectives. Nevertheless, the epistemological basis of the construct needs to evaluated. The next chapter will be dedicated to that issue.

3
The Epistemology of Creativity

Introduction

In this chapter, the notion of organizational creativity will be examined in terms of its epistemological underpinning. While the literature on organizational creativity in many cases simply assumes that there is something called organizational creativity, closely associated with the creation of new ideas or solutions to problems, the aim of this chapter is to critically engage with the assumptions on which the construct of organizational creativity are based. For instance, there is a rather pervasive belief in the literature that organizational creativity is dependent upon the cognitive and imaginary capabilities of individual human beings – the creative individual – serving as the smallest unit of analysis. But this image of the creative human is far from being value-neutral or self-evident; instead, it draws on a number of ideologies giving priority to the individual human subject at the expense of the collective and postulating a (philosophically) humanist explanation wherein objects such as technologies and non-humans (to employ the actor–network theory vocabulary) are eliminated from the analysis. As a consequence, the dependence upon the individual human agent in the explanatory framework reduces a rather heterogeneous network of relations between humans, technology, laboratory equipment, information systems, and so forth, to the level of the individual – an anthropocentric view of organizational creativity. Not only does this reduction rest on frail epistemological grounds, but it also provides an unnecessarily simplified model of how organizational creativity works in practice. In terms of management practice, an overly narrow view of creative work provides us with few opportunities to actively support and reinforce the creative activities in organizations. As a consequence, invoking notions such as ontology and epistemology may appear

Table 3.1 Two epistemologies of creativity

Mainstream view	Alternative view
Subject-centred creativity	Distributed creativity
Creativity as discrete event	Creativity as continuous and connective event
Social constructivist view of creativity	Material (biologically true) and social constructivist views of creativity combined

somewhat detached from the day-to-day activities of organizations, but there are direct positive practical effects from such an analysis in terms of clarifying the relationships between different resources in organizations.

In this chapter, three different epistemological perspectives on organizational creativity will be discussed as contrasts against the mainstream view of organizational creativity (see Table 3.1).

In the first section, the subject-centred view of organizational creativity is criticized and a more distributed image of organizational creativity is examined. In the second section, organizational creativity as a discrete event is contrasted against a continuous and connectivistic view of organizational creativity. In the third section, the social constructivist view of organizational creativity is debated and compared with a materialist definition of organizational creativity, drawing on the biological effects of a certain innovation in the field of biomedical sciences. These different epistemological perspectives offers a more pluralist view of what organizational creativity is and how it works.

The notion of creation

We start out with the lexical, 'congealed' and 'non-creative' or even 'a-creative' definition of creativity (see Bataille, 1983; Castoriadis, 1987) in order to establish a shared ground for a discussion of the concept. Webster's Dictionary defines creativity as the 'ability to create' and advances 'uncreativeness' as its antithesis. Creativity is, in turn, defined as 'having the ability or power to create' and as being 'characterized by originality and expressiveness', as in 'creative writing.' Next, creation is defined as 'the act of creating', or 'the fact or state of being created'. These definitions indicate that creativity is grounded in itself, as being either an *act* ('to create') or *a quality* ('creative solutions', 'creative thinking') rather than being based on two or more external processes or as signifying some underlying activities. Whitehead (1927) points out that the English notion of creativity is etymologically derived from the Latin *creare*, 'to bring forth, beget, produce'. Thus, creativity is about producing new things, ideas, or entities. Therefore, to Whitehead (1927: 21), 'creativity is the principle of *novelty*'. David Bohm, an English physicist praised as a major contributor to the field of quantum physics (see Lucas, 1989), has discussed creativity in a number of books (see for instance, Bohm, 1998; Bohm and Peat, 1989). To Bohm, creativity is not something that can be fully planned and controlled. He writes: '[o]riginality and creativity begin to emerge, not as something that is the result of an effects to achieve a planned and formulated goal, bur rather as a by-product of a mind that is coming to a more nearly normal order of operation' (Bohm, 1998: 26). In addition, creativity is for Bohm a mark of originality, rather than one of superior intelligence or diligence. Bohm argues: 'There are a tremendous number of highly talented people who remain mediocre. Thus, there must have been a considerable body of scientists who where better at mathematics and know more about physics than Einstein did. The difference was that Einstein had a certain quality and *originality*' (Bohm, 1998: 3). According to this view, creativity is the outcome from original thinking based on the will to develop new ideas rather than to conform what is taken for granted or commonly shared knowledge. Creativity is what uproots the taken-for-granted beliefs and ideas within a community. In other words, creativity is the ability to provide new ideas, thoughts, artifacts, images, and so forth, that radically breaks with what was previously accepted as legitimate truths or conventions. In Whitehead's (1933: 179) words, 'the creativity is the actualisation of potentiality, and the process of actualisation is an occasion of experiencing'. Creativity is a form of becoming, a movement from what is possible to what is actual; creativity is always taking place in the borderlines of what is known and familiar. As a consequence, creativity is never self-assured because it is not operating in realms that are what Giddens (1990) calls *ontologically certain*, that is, the creative human being cannot know if the activities are creative *ex ante* but must trust intuition and gut feelings and other cognitive and emotional faculties that extend outside the domain of instrumental rationality. In other words, creativity is a most complex ontological and epistemological construct, operating in the realm of being 'in-between' the known and the unknown, the familiar and what is becoming, in a state of liminality. Thus, the need for an elaborated theoretical analysis of the construct.

Creativity and the individual

In psychological research programmes exploring creativity there is a long-standing tradition to examine individual human beings and their creative capacities. The British psychologist Winnicott (1971: 65) has emphasized the notion of creativity as one of the most important existential qualities of human lives:

> It is creative apperception more than anything else that makes the individual feel that life is worth living. Contrasted with this is a relationship to external reality which is one of compliance, the world and its details being recognized but only as something to be fitted in with or demanding adaptation. Compliance carries with it a sense of futility for the individual and is associated with the idea that nothing matters and that life is not worth living. In a tantalizing way many individuals have experienced just enough of creative living to recognize that for most of their time they are living uncreatively, as if caught up in the creativity of someone else, or of a machine.

For Winnicott (1971), creativity is a form of playing, a sense of masterfully controlling one's own life world. The Dutch historian Johan Huizinga (1949) gives the role of playing a similar role in culture and human experience; playing and creativity are closely interrelated. In almost all texts on creativity, the act of creation is treated as something that is both gratifying and highly appreciated. This is not to say that creativity does not demand substantial degrees of concentration and hard work or that there is no sense of frustration intrinsic to creative work. In most cases, creative work places immense demands on the individual. Yet the sense of 'breaking through' cognitive barriers and reaching an understanding of a certain matter is one of the most intense and profound human experiences. On the level of psychology and psychoanalysis and culture sociology, the notion of creativity and creation is adequately conceived of as individual acts or at best acts that happen in association with other human beings. In such accounts it is the individual who is or becomes creative through his or her own engagement with a set of cognitive, emotional, or perceptual human faculties. Research programmes that take their point of departure in the individual human being offer a specific form of theory about creativity, very much determined by the implicit assumptions of the paradigm. However, psychological research programmes are not always of immediate practical consequences for managers and co-workers in organizations and companies. Creativity in laboratory settings and in real-life situations may occur in a rather different manner. Nevertheless, creativity remains in many cases associated with the individual human subject. The implicit epistemological assumptions underlying to this view are rarely addressed in the creativity literature.

In French post-Second World War philosophy and thinking, the notion of post-humanism or anti-humanism (Kallinikos, 1998; Willmott, 1998) has been invoked when formulating a critique of the Cartesian subject being separated into a cognitive or thinking substance (*res cognitans*) and an embodied and fleshy substance (*res extensa*) (Gatens, 1996). The cognitive substance is not extended and is representing the soul in the Judeo-Christian tradition while the embodied and extended substance represents the body. The Cartesian operation of distinguishing the mind and the matter, the body and the soul, has been the dominant doctrine in the western canon. Still, this philosophical system had always been contested. One of the first prominent critics of Descartes was in fact Spinoza (1994). In their writings post-structuralist and postmodernist scholars and philosophers such as Michel Foucault, Jacques Derrida, Jean-François Lyotard and Gilles Deleuze have offered a critique of the reductionist program of Descartes. Derrida is talking about the Cartesian split as a form of *logocentrism*, what Heidegger terms the *philosophy of presence*, wherein the subject is located in a single point and where reason and a number of human faculties are inscribed into the human body. In Foucault's account, the Cartesian humanism represents a specific form of thinking that emerged with the Renaissance and that emphasized the human as the locus of reason. For Foucault, this image of man is, however, dependent on specific historical conditions that may alter as new forms of knowledge emerge. Lyotard has examined the humanist programme as a form of grand narrative, an explanatory story aimed at locating reason and rationality in man. It is becoming increasingly complicated to maintain this narrative as a legitimate and credible explanation since there is too much evidence of irrational behaviour in human actions. Deleuze, finally, has formulated a substantial critique of the Cartesian programme as a thread of thinking that runs from Plato through Descartes and beyond and that may be de-familiarized through the use of alternative philosophical resources such as Spinoza, Nietzsche and Bergson. A shared theme in French post-structuralist thinking is the scepticism towards the idea that it is the individual single human being that is the sole representative of reason and rationality. Speaking in terms of creativity and the management of creativity, the implicit and highly epistemological assumption that it is the individual that is the primary locus of creativity becomes a contested one. There are a number of philosophical programmes and research programmes that offer a critical view of this humanist and subject-centred model of creativity. For instance, the literature on Science and Technology Studies (STS), the sociological analysis of scientific production in laboratory settings and other scientific practices, offers detailed examination of how practicing scientists work in their day-to-day life (Jasanoff *et al.*, 1995). Among other things, the STS literature emphasizes the co-dependency between the individual researcher, the theoretical framework guiding the operations, the technical equipment and tools, and various non-human resources (Knorr Cetina, 1999; Traweek, 1988; Latour, 1987; Lynch, 1985). In other words, there is an image of scientific work as being a highly distributed practice, integrating a variety of heterogeneous resources that are jointly adjusting to one another. Pickering (1995) offers a detailed analysis of what he calls the 'mangle of practice', the continuous change and movement taking place between the various components in laboratory work. In Pickering's account, the machines being used in research are not just mere tools, but are given an epistemological status in-between human and the non-human (the object of study). Pickering (1995: 7) writes: 'The machine, as I conceive it, is the balance point, liminal between the human and nonhuman worlds (and liminal too, between the worlds of science, technology, and society)'. Since scientific work is an outcome from the ability to effectively coalign humans, technology and non-humans in the course of action, Pickering is moving beyond the humanist terrain giving the full prerogative to do science to the human subject. Instead, Pickering is talking about a 'posthuman space', that is, 'a space in which the human actors are still there but now inextricably entangled with the nonhuman, no longer at the center of action and calling the shots' (Pickering, 1995: 26). Science is not solely an effect of the human's cognitive faculties, but is rather a combination of resources that may provide a series of 'empirical statements about the world' (Pickering, 1995: 68). Such 'empirical statements' therefore constitute a theory or a proposition when being evaluated by peers in the research community. Pickering (1995: 70) concludes: '[S]cientific knowledge should be understood as sustained by, and as part of, interactive stabilizations situated in a multiple and heterogeneous space of machines, instruments, conceptual structures, disciplined practices, social actors and their relations, and so forth. This is my version of Serres's idea that "nature is formed by linkings".'

In a humanist, subject-centred epistemology, creativity is what is safely located within the human subject; his or her abilities to conceive of new ideas and new statement on the basis of empirical investigations is only *supported* by the use of technology and tools. In the post-human framework advocated by Pickering (1995), it is not useful to give priority to the human since it is the system of interrelated resources and its mutual adjustment that is serving as the foundation for any scientific contribution – that is, any creative solution to paradigmatic problems. In terms of creativity, Pickering's study offers an alternative view of the notion of creativity. Rather than staying within the realm of the subject, theories on creativity may move into new epistemological domains.

Creativity as process

The next underlying assumption in the literature on creativity is to conceive of creative acts as being instantaneous and appearing in a single moment. The mythology of scientific findings often credit major scientific breakthroughs with the quality of being single, isolated events. This folklore of science postulated that scientific findings are like lightning strikes, once the time has come and the information has been processed long enough in the scientist's or artist's mind. In real-life settings and in laboratory work, as, for instance, in the case of new drug development in the pharmaceutical industry, there are few such points of discovery; rather the day-to-day work emerges along series of events structured into standardized laboratory practices that are moving towards more detailed investigations. Such series of events do not generally proceed along linear paths, but may be unpredictable. Isolating creative findings into discrete points represents what Whitehead (1927) calls 'the fallacy of misplaced concreteness', that is, the fallacy of locating particular qualities or occurrences within specific isolated entities or events (see Chapter 7 for a more detailed discussion of this point). Creativity does not occur at a single point in time but is rather, one may argue, the outcome of a series of interconnected events and undertakings. In this section, the notion of creativity will be examined as a form of *connectivity*, the ability to make connections between heterogeneous materials. In this view, the notion of creativity escapes the misplaced concreteness postulated by the mythology of creativity. In the following, the philosophy of the French thinker Gilles Deleuze will be invoked. In more specific terms, Deleuze's notion of the *rhizome* – that is, a model of knowledge that is horizontally dispersed in a single plane rather than in the commonplace tree structure emerging along different paths separated from one another – will be discussed as a fruitful image of creativity.

Deleuze and the notion of rhizome

In this section, the concept of organizational creativity and its various practices are discussed in the framework of the thinking of the French post-structuralist philosopher Gilles Deleuze.[14] In contrast to the subject-centred perspective of creativity, the concept of organizational creativity is used to acknowledge the context-specific and collaborative aspects of creative acts in organizations. Before we engage in an analysis of creativity, the thinking of Deleuze needs to be introduced. Deleuze is one of the most important post-Second World War thinkers and can be categorized as a post-structuralist philosopher (Best and Kellner, 1991; Ansell Pearson, 1999). To date, Deleuze has been only marginally recognized within social sciences and organization theory, but within the humanities there is an awakening interest in Deleuze's complex philosophy. For instance, in the latter half of the 1990s, a number of introductory books and articles on Deleuze have been published (Badiou, 1999; Marks, 1998; Stivale, 1998; Hayden, 1998). Badiou (1999: 96) writes about Deleuze:

> He was neither a phenomenologist not a structuralist, neither a Heideggerian nor an importer of Anglo-American analytic 'philosophy,' not again a liberal (or neo-Kantian neohumanist)... As with all great philosophers, and in perfect conformity with the aristocraticism of his thought and his Nietzschean principles of the evaluation of active force, Deleuze constitutes a polarity *all by himself.*

Deleuze was a highly original thinker, making use of his own favourite philosophers, such as Spinoza, Leibniz, Nietzsche and Bergson, in order to create new opportunities for thinking. Hayden (1998) remarks:

> Although Bergson, Nietzsche, and Spinoza are radically different thinkers whose philosophies are often vastly divergent, for Deleuze they are all united on these points at least: The critique of transcendental realms, causes, values, and principles, and the affirmation of a dynamic, fluid, and immanent world within which human beings exist and create diverse ways of living. In this respect, all three thinkers are regarded by Deleuze as belonging to a philosophical

---

[14] In this chapter we make use of Deleuze's name as a synecdoche for the joint co-authorship between Deleuze and Félix Guattari. As we refer to Deleuze's other philosophical writing not co-authored with Guattari in the paper as well as to Deleuze and Guattari's joint publications, we have, for the sake of simplicity, been talking about *'Deleuze's* model of the rhizome'. This does not mean that Guattari's contribution is not recognized. In fact, Guattari's contribution to the four co-authored books of Deleuze and Guattari is every bit as substantial as Deleuze's.

tradition that affirms immanence and criticizes supernatural, divine, or mythical versions of transcendence. (Hayden, 1998: 68–9)

Deleuze breaks with a Platonist tradition of thought in western thinking and embraces a process-based, non-transcendental tradition of thought promoted and represented by, for instance, Bergson and Whitehead in twentieth-century philosophy. In social sciences, Deleuze has been referred to by feminist theorists such as Braidotti (1994, 1997), Grosz (1994, 1995), and Olkowski (1999), and cultural and post-colonial theorists (e.g., Buchanan, 1997; Young, 2001). In organization studies, Deleuze has been employed in finance (Bay, 1998), accounting (Bougen and Young, 2000), organization change studies (Chia, 1999) and human resource management (Brewis and Linstead, 2000). In this chapter, the notion of the rhizome, a conceptual model aiming at breaking with a Platonist tradition of thinking offered by Deleuze, will be invoked as a tool when studying creativity in organizations.

In their massive volume *A Thousand Plateaus* (1988), Deleuze and his co-author Félix Guattari, a French psychoanalyst and social theorist, develop the idea of a interconnected network or field that breaks radically with the tree metaphor of knowledge that prevails in western thinking (see, for example, Maturana and Varela, 1992). In a commentary on Deleuze's thinking, John Marks (1998: 45) writes:

> The rhizome is a figure borrowed from biology, opposed to the principle of foundation and origin which is embodied in the figure of the tree. The model of the tree is hierarchical and centralized, whereas the rhizome is proliferating and serial, functioning by means of the principles of connection and heterogeneity.

While the tree model of knowledge implies that all branches of the tree can be located back to the roots and stem, the rhizome model operates on a single plane, a very important image of reality that Deleuze borrows from Spinoza (Deleuze, 1988a: 122). To Spinoza, a prominent critic of Descartes, the world is emerging in a single plane, in a plane of immanence (Spinoza, 1994; Deleuze, 1988a, 1990). All events and entities are produced from the same substance. To Bergson (1911), another thinker who had a considerable influence on Deleuze, being is in a restless state of becoming. Nature and society is continuously altering itself and new species and events are continually produced. Both Spinoza and Bergson share with Deleuze the view that there is no such a thing as Platonist transcendental knowledge that we can take part of through various activities. To Spinoza, Bergson and Deleuze, there is only one single plane of existence wherein new events and entities are produced. Patton (2001: 1094) writes: '[t]he rhizome stands for a non-hierarchical, a-centred field of knowledge. It stands for multiplicity as opposed to the principle of unity, and for open-ended creation of new ideas as opposed to the reproduction or repetition of established patterns'. Rather than seeing knowledge and thinking as always already being related to one single or a number of sources, the rhizome operates *horizontally*. Deleuze and Guattari (1988: 7) talk of this as the 'principles of connection and heterogeneity' meaning that 'any point of a rhizome can be connected to anything other, and must be'. They write:

> [U]nlike trees or their roots, the rhizome connects any point to any other point, and its traits are not necessarily linked to traits of the same nature; it brings into play very different regimes of signs, and even nonsign states. The rhizome is reducible neither to the One nor the multiple. (Deleuze and Guattari, 1988: 21)

As a consequence, a rhizome is, in contrast to the tree model, not based on the ability to trace all knowledge and statements back to the roots, i.e. their 'origin'. Thus, the rhizome is an 'anti-genealogy', it does not assume that knowledge are strictly related to a single set of influences, but that knowledge and statements appear whenever there are connections made within the rhizome. Ansell Pearson (1999) writes: 'The rhizome is "anti-genealogy" since it operates in the "middle" without *arche* or *telos*, operating not through filiation or descent but via "variation, expansion, conquest, capture, offshoots"' (Ansell Pearson, 1999: 158). Therefore, a rhizome can never claim to offer any transcendental truths; the rhizome only gives a number of connections that in turn produce new opportunities for action. 'A rhizome has no beginning or end; it is always in the middle, between things, interbeing, *intermezzo*' (Deleuze and Guattari, 1988: 25). In a tree model of knowledge, all new knowledge could be immediately referred or traced back to a set of influences. Thus, there is a distinction made between 'new knowledge' and 'old knowledge', between classic and new ideas. This model of thinking is the traditional Platonist image of knowledge that Deleuze and Guattari reject. In a rhizome, there is no one single influence, tradition, historical programme, and so forth that could be used to legitimize an idea. In a rhizome there is no 'genealogy'. There is nothing but series of connections, 'lines of thoughts' that are developed in a single plane. The tree model is hierarchical, meaning that some ideas are prior to others or more established in terms of legitimacy. In Deleuze and Guattari's thinking, such conservative images of thought cannot be maintained because they presuppose that all knowledge can be traced back to certain roots. In a rhizome, one would speak of *routes* rather than roots (see Clifford, 1997). Thus, the rhizome offers unlimited opportunities for connections and therefore for the creation of new ideas. Commentaries of Deleuze's thinking make at times use of the Internet as being an illustrative model of a rhizome (see Poster, 2001; Stivale, 1998). The Internet is not organized in accordance with a tree model: It is a network of independent servers and websites that can be interconnected to one another. In the Internet, there is no 'master-website' or centre that determines the relationship between the other homepages. Any homepage could be related ('linked') to any other homepage with an adequate homepage address. The Internet is thus a rhizome, a horizontal multiplicity wherein all entities can be related to one another.

The rhizome model could be used as a fruitful model for organizational creativity because of its emphasis on a free play of the resources within the rhizome. The tree model of knowledge is more conservative because it does not acknowledge the creative force of new ideas. The most ideal-typical model of knowledge based on the tree model is Plato's philosophy of knowledge, wherein truth and opinions, knowledge and belief, are clearly distinguished. To Plato, knowledge is based on a recollection of eternal, transcendental ideas accessible for the individual through philosophical training. Deleuze and Guattari reject any image of knowledge that is based on the Platonist idea of recollection. To Deleuze and Guattari, thinking is always creative, immanent, a force enabling new ideas to emerge. Thus, the rhizome model of knowledge is anti-genealogical and emphasizes the series of thoughts produced through connections. To Deleuze and Guattari (1988), the rhizome represents a new image of thought that radically breaks with the Platonist tradition of thinking.

New drug development in the pharmaceutical industry: creativity as a rhizome

The pharmaceutical industry is fundamentally based on the production of new drugs offering valuable therapeutic effects for its end users. New product development in the pharmaceutical industry is basically composed of four phases in the discovery and development organizations. It is here noteworthy that what is referred to as *drug discovery* onwards is primarily denoting the formal organization of the pharmaceutical company's organization. Speaking from within a Deleuzian framework, it would be more adequate to talk about discoveries as 'events' wherein certain resources are connected to one another. Thus, the use of the notion of 'discovery' does not imply a foundationalist view wherein things *per se* are unconcealed.

The research process in the pharmaceutical industry is long and complex; it is a major undertaking that often runs over 15 years. It involves many scientific disciplines and technologies. On average, it costs a company more than US$500 million to get one new medicine from the laboratory to the pharmacist's shelf (Thompson, 2001). Development of a pharmaceutical product can be generalized by dividing the research into three major processes: (i) discovery (ii) development, and (iii) product support and life-cycle management. Figure 3.1 provides a brief overview of the process and the involved disciplines.

The primary objective for discovery is to identify new molecules with potential for producing a desired change in a biological system (e.g., to inhibit or stimulate an important enzyme,[15] to alter a metabolic pathway, or to change cellular structure) (Hullman, 2000). The drug discovery process, phase 0, begins by defining a disease area and a target to manipulate. This process may require, for example, research on

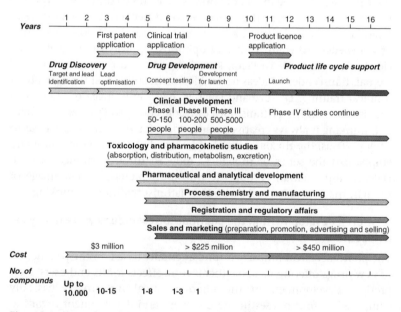

Figure 3.1 Overview of the drug research process[16]

---

[15] An enzyme is a class of large proteins, which can catalyze a broad spectrum of biochemical reactions; it is formed in living cells. Enzymes contain polypeptide chains with a molecular weight that ranges from 10,000 to 1,000,000.

[16] AstraZeneca *Annual Report*, 2001.

fundamental mechanisms of disease or biological processes, research on the action of known therapeutic agents, or random selection and broad biological screening. New molecules can be produced through synthesis or extracted from natural sources (plant, mineral, or animal). The number of compounds that can be produced, based on the same general chemical structure, runs into the hundreds of thousands (Jones, 2001).

The target may be a receptor, for example, or an enzyme. The medicinal chemists synthesize substances that are tested for activity in relevant *in vitro* systems or biological models by biochemists and pharmacologists (Lesko *et al.*, 2000). These tests involve use of animals, isolated cell cultures and tissues, enzymes and cloned receptor sites and computer models. If the results of the tests suggest potential beneficial activity, related compounds (each a unique structural modification of the original) are tested to see which version of the molecule produces the highest level of pharmacological activity and demonstrates the most thera-peutic promise, with the smallest number of potentially harmful biological properties (Gregg, 1997). The aim is to establish a chemical structure for the biological activity relationship, which in the successful project leads to a candidate drug (CD).[17] The CD is then further tested for putative toxicity and, if found safe, an application (investigation of a new drug [IND]) is filed with drug regulatory authorities and ethical committees – to obtain approval for testing on humans. The discovery process is complex and unpredictable and involves many factors that could influence the successful outcome. It normally takes around three to five years to produce a CD.

After authorities approve the IND, clinical studies can then begin. The required three-part clinical trials process, which judges the efficacy and safety of potential treatment, is a major undertaking. The first studies (phase 1) are started in humans, usually using healthy volunteers. The aim is to study tolerability of the drug and its pharmaco-dynamic and pharmaco-kinetic properties (i.e., seeing how the drug affects the body and how the drug is affected by the body, respectively). If possible, the dose effect and time effect relationship is studied in phase 1. The aim of the clinical studies is screening for safety, which means gathering information on whether the drug is safe to give to humans and, if so, how much they can tolerate. Phase 1 studies aim to find the appropriate

---

[17] A candidate drug exists in the final pre-clinical stage of drug development, which denotes the selection of a compound with the greatest potential to be developed into safe, effective medicines.

dose range. Of course, before human testing begins, the general safety of the drug is established in animals. These studies normally take one to two years to perform. Then the new drug is administered to patients for the first time in phase 2.

The main goal of phase 2 testing is pragmatic: to find experimental conditions to establish an optimal dosing regimen and in particular, establish results of primary endpoints, which describe unambiguous results that indicate exactly what the treatment can do (Friedman, Furberg and DeMets, 1985). So phase 2 studies clarify whether the drug has the desired therapeutic effect and dose effect relationship (i.e., proof of concept). This phase marks the introduction of the control group to the trial.[18] Phase 3 studies constitute the final clinical trial stage. Here, the goal is to establish the role and documentation of the new drug in the current state-of-the-art therapy arsenal. These studies are large scale – thousands, or even tens of thousands, of patients participate (Zivin, 2000). By this point, researchers who are running the trial have defined at least one group of patients that is expected to benefit, and the best way to administer treatment. If, after careful statistical analysis, the candidate drug proves to be significantly more effective than the control treatment, the trial is called pivotal. Ordinarily, two pivotal studies are needed to prove the new therapy's value. These studies examine the effect of the new drug when compared with reference substances and often, with a placebo. The typical length of these studies is three to four years. After completion of phase 3 studies, the final documentation can be compiled and submitted to the appropriate national regulatory agencies (e.g., the FDA) for review (new drug application [NDA]); Hullman, 2000). After approval, the product can be marketed. Adverse effects are followed meticulously through all clinical phases and after approval of the drug for launch. If the candidate drug is approved, the clinical effects of the drug are studied further in phase 4 studies, which can be quite extensive; they often take four to six years to perform. The clinical research programme continues after the product's launch – by collecting data from outcome research and epidemiology data from patients; this might lead to new indications for the product (Zivin, 2000).

One of the most important phases in the new drug development process is the discovery. The key process is the identification and

---

[18] Ideally these studies are double blind, which means that neither physicians nor patients know whether a subject is part of the treatment group or the control group.

synthesis of NCEs[19] (the potential precursor of a CD) and the clinical verification of its therapeutic effects. Each NCE is a new molecule that affects biological organisms in a benevolent manner. Within the discovery phase there is a continuous alteration between laboratory findings (e.g. chemical structural relationships, a chemical entity) and its biological verification on an organisms. In the discovery phase, the basic and applied research activities are being bridged: on the one hand, there is the synthesis of the NCE, and on the other hand, there is the experimentation on the effects of the NCE *in vivo*. Therefore, the discovery is based on the ability to make connections between different entities. First, there are the connections between chemical substances on a molecular level enabling for the establishment of new chemical entities. For example, medicinal chemists, pharmacologists and biologists are actively producing new connections between chemical entities and target profiles in order to produce new configurations. Secondly, there is the connection between the NCE and the organisms. In this case, the interaction between NCE and the organism must give rise to some efficient and safe therapeutic effects. If the NCE only offers a marginal effect of the targeted indication, for instance the effect of reducing blood pressure, it may be abandoned in order to identify another NCE that does give the desirable effect. If the NCE and the organism are producing the desired effects, it may be selected as a CD and is further tested in complementary models and, later, on voluntary human beings. In the discovery phase there are thus a number of connections in a single plane that constitute creative solutions to practical problems in terms of health and well-being. First, the NCE is a new configuration, a new molecule structure that may be proven to offer desirable effects. Secondly, the connections between the NCE and the animal organisms must prove to be viable and to offer significant effects, or else the NCE will be abandoned. The early phases of the new drug development activities are thus based on the establishment of lines of thought and connections in a single plane.

In the second phase, a NCE that have been proven to offer significant results and are chosen to become one of the targeted new drugs of the pharmaceutical company is being developed. Drug development comprises all the activities aiming at establishing the qualities of new drug, ranging from safety, tolerability, efficacy to having a proper

---

[19] A new chemical entity is a compound that is not previously described in the literature.

administration route and control of active substance in the final drug formulation. In the discovery phase, medicine and pharmacology become important disciplines offering expertise knowledge on the human body's capacity to absorb the NCE. The development phase include a number of issues to be addressed in terms of immediate bodily effects on the patients. In the discovery phase, it is primarily chemistry and biology that are the prime movers, but in development it is often medicine and pharmacology. In the early phases, there is no immediate involvement from human organisms. The NCE are tested on animals. In the development phase, human beings are using the CD in order to verify and validate its therapeutic effects. As a consequence, the new drug development process is being composed of lines of connections across disciplines and practices: chemistry is connected to biology that is further connected to medicine and pharmacology; the NCE is tested on animals and later on human beings and become a CD and is finally turned into an object of clinical research. The point of departure is, with Knorr Cetina's (1999) formulation, 'a science of life without nature', a laboratory activity aimed at discovering a new chemical entity, but, as the new drug development process proceeds, the NCE is being turned into a CD and in the end a new registered drug. Thus, the entire new drug development process is starting in a laboratory as a vision of a new chemical substance being able to affect, say, blood pressure in a human organism, and is finalized as a new drug being tested on thousands of patients across the globe. The process is passing on from the initial states of belief or vision into the stage where the new registered drug is seen as a 'fact' in terms of its ability to offer therapeutic effects (cf. Latour, 1987).

One way to examine these lines of activities, the various phases wherein the NCE or CD is being passed on like a token, is to examine the connections between the various entities that are being invoked and mobilized throughout the new drug development process. To speak with Deleuze, we can say that the new drug development is being based on a structure that resembles the rhizome model of knowledge. In a rhizome, all nodes can be connected to one another. The series of connections make up a line of thought, a trajectory in which one single idea is being developed. In pharmaceutical research, one such idea is the search for a new drug with desirable therapeutic effects. As that idea comes into being – the becoming of a new drug – various connections are being made: the synthesis chemist is making connections on a molecular level in order to identify the NCE sought for; the chemist is later making connections with the biologist to verify the qualities of the NCE in relation to an organism; the NCE is connected to the human organism, and so forth. The idea of a new cardiovascular medicine is proceeding through its various connections within this rhizome-like structure. All connections are aimed at solving practical problems and, therefore, the activities are taking place on a single plane wherein chemical substances, animals, human beings, researchers and patients are being connected to one another. When making use of Deleuze's rhizome model when examining creative activities, creativity unfolds as a series of activities and connections. Creativity is not removed from everyday life practices and events, but is continuously producing effects on the single plane. To Deleuze, knowledge and ideas cannot be traced back to the roots of the tree ('the rhizome is an anti-genealogy') but is always based on connections, on connectivity, on the ability to establish lines of thoughts between entities and events. In the case of new drug development, the initial event of the 'discovery' of the NCE is later turned into a multiplicity of events as the NCE, is leaving the laboratory setting and finally encounters the thousands of patients in full-blown clinical trails. The NCE is passed on from the laboratory into the human society via the biological models verifying its status as a major component of a CD.

In the Deleuzian conceptualization, creativity is the ability to make connections. Creativity is neither external to the individual in terms of being determined by 'work climate' or 'shared worldviews within communities', nor is it the supreme quality of a minor number of highly skilled, talented or extraordinary human beings. Creativity is neither contingent nor based on elitism; it is *connectivistic*. Connections are what make a difference, they are enable new ideas and new entities and events to occur. The rhizome is an ideal-typical model for how knowledge is based on immanence – on innate relations rather than transcendental truths. As being a ideal-typical ontological and epistemological model it is applicable in various cases. When examining and theorizing creativity, the rhizome model offers opportunities to show how creativity is based on connection and lines of thoughts across an horizontal structure. The rhizome model therefore acknowledges that creativity neither falls from the sky, nor emerges from extraordinary conditions. Therefore, the rhizome model helps us to demystify creative processes. In the light of this model, creativity is not transcendental but immanent; it is produced *within* the series of connections rather than influencing the connections.

Benefits of the connectivity perspective

This section has aimed to offer one possible model for creative thinking: Deleuze's model of the rhizome. The rhizome is a horizontal structure that is based on its connectivity, that is, its ability to connect its various nodes in a multiplicity of combination. To Deleuze, the rhizome is an ontological model, informed by the philosophy of Spinoza and Bergson, that breaks with the Platonist ideas of presence and origin, in short the tree image of knowledge. The rhizome is thus to be seen as an ontological and epistemological model as well as a tool for analysis. It is of a most abstract nature but it does offer new opportunities for thinking, new ideas, and new conceptualizations. Within the rhizome, all sorts of syntheses and connections are possible. Just as the Internet provides a multiplicity of opportunities for connections, pharmaceutical research comes into being through lines of thoughts constituted by a multitude of connections. However, this does not, mean that every connection is equally viable. The rhizome offers possibilities for the creation of new associations and connections, but only a small number of these new entities and events will prove to be useful and viable. In comparison to pharmaceutical research, a great variety of connections and syntheses are possible but only a fraction of these are useful and viable, i.e., will prove to have benevolent and safety affects for the human organism.

The rhizome is a non-reductionist model of creativity. It underscores that what we are seeing as creative solutions and creative ideas are always produced through association across various entities and events. Creativity is the line of thought that emerges when connections are being made. To Deleuze, the image of the rhizome is of great importance as a model that radically breaks with Platonist metaphysics, the dominant tradition of thoughts within western philosophy. Instead of considering knowledge and creativity to be dependent on transcendental truths and ideas, Deleuze claims that what is of interest is not of extra-social origin but that everything interesting happens 'in the middle', among the turmoil of everyday life, in between what is taken for granted and what is seen as legitimate knowledge (Deleuze, 1997: 2). To Deleuze and Guattari (1988: 21) the rhizome is 'in the middle', it is where things take place and where new connections are being made. Taken together, Deleuze presents a alternative model of thinking that is based on pre-Socratic thinkers such as Heraclitus and Parmenides, Roman philosophers such as Lucretius, so-called 'new philosophers' such as Spinoza and Leibniz, and modern thinkers such as Bergson (Hayden, 1998; Braidotti, 1997). To Deleuze, the world is immanent. But this does not imply an Aristotelian *entelechism* (see Aristotle, 1986) in which all entities are determined by their own potentiality. For Deleuze, the world is never determined, it is always unfolding as an opportunity for new connections and new syntheses. Therefore, Deleuze is a philosopher whose thinking can be located within a long tradition of thought at the same time as he respresent a radically new and highly idiosyncratic philosophy (Surin, 1997). Following Badiou (1999) we may conclude that Deleuze's thinking is always on the move, always in a state of becoming, just as his admired forerunners Spinoza, Leibniz and Bergson's thinking represents a philosophy of becoming (see Deleuze, 1988a, 1988b, 1990, 1993).

From a practitioner's point of view, the implications from the conceptualization of creativity in terms of being a rhizome are that creativity can be managed in terms of making connections possible. To Deleuze, the message is clear: creativity is what emerges in rhizome structures as lines of ideas and thought. The major event in the rhizome model is the *connection* when one single idea is related to another idea and another synthesis is produced. The acting manager who wants to reinforce creativity needs to ensure that creativity is being emphasized throughout the organization. In the case of the pharmaceutical industry, it is evident that the close cooperation between, for example, chemists, biologists, physicians, and pharmacologists enable new ideas to emerge and materialize as new developed drugs. Throughout the process, numerous connections are being made; the chemical substance is created and is verified in relation to the biological model. Next, the NCE is clinically tested on a human organism. Finally, it is verified on large-scale clinical trails. The pharmaceutical researchers are here connecting to one another. They pass around information relevant to the NCE in various communities. They jointly make sense out of the 'chemical substance' through its association with different communities. The entire new drug development process is producing a line of thought across the rhizome network in which the new drug development takes place. There are no transcendental truths inherent to a specific new drug, only a great number of connections and associations on a single plane. From the practitioner's point of view, organizational creativity can be managed through transparency and visibility. To promote a continuous exchange and dialogue in organizations is of great assistance to individuals who want to come up with creative solutions and new ideas. This can be done, for example, by stimulating and rewarding networking activities and improve the sharing of scientific information across the organization. A major idea of the rhizome model of creativity is that there is nothing mysterious about creativity; it is neither a myth, nor a extra-social or extraordinary quality possible only for a few individuals. Creativity is an outcome from social practices and social interaction within the plane of resources and relations (Styhre and Sundgren, 2003a).

In the rhizome model, creation and creativity is an event in which new connections are made in a single plane. In the end, the new chemical entity can be turned into a new registered drug as an effect of successful connections being made. The pharmaceutical research field is therefore a rhizome in which various actors, processes, events, entities, etc. are being continuously connected, de-connected and re-connected. From a practical point of view, there is a multiplicity of opportunities to enable for more creative activities in terms of offering more possibilities for communication and exchanges within organizations. What is at stake in the rhizome network is full control over the process, but what is gained is a more dynamic, more progressive and creative organization.

Social constructivist and material definitions of creativity

Ford (1995a) and Csikszentmihalyi (1999) argue that creativity is not an inherent quality of a person, process, product or place, but is rather a domain-specific social construction that is legitimized by judges who serve as gatekeepers to a particular domain. This perspective becomes particularly relevant to many organizations, such as those in the media or advertising industries, architecture, and IT design. Here, 'creative contributions' are closely intertwined with social constructions, such as mechanisms that attract attention and interest. In comparison to these industries, the pharmaceutical industry is based on scientific work – at least partially separated from social constructions, such as public opinion and common beliefs. But for the pharmaceutical industry and the life science industry, this perspective must be adjusted. The pharmaceutical industry, in particular, serves as an interesting example for demonstrating an intermediate role between hard sciences (such as biology, medicine, pharmacology, and biochemistry) and the market.

Creativity in new drug development is under the influence of two major factors: regulations and scientific breakthroughs. From an epistemological perspective, one may divide creativity into two perspectives (see Figure 3.2). The first perspective of creativity consists of an inner *core* of hard realism that reflects objective truth.[20] The inner core is then

---

[20] Truth is a philosophical concept of great difficulty. Here *truth* means the fact that a certain substance shows predictable responses in the human organism. For example, the drug product Losec/Prilosec has an effect on certain receptors that cause a decrease of acid secretion in the stomach, which allows, for example, the more rapid healing of gastric ulcers. Truth is thus denoting the ability to predict that a substance has a significant biological effect.

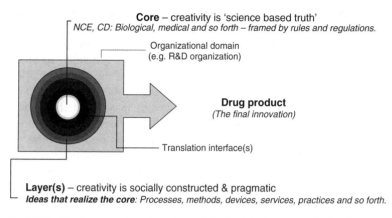

Figure 3.2 Dual perspective view of creativity in the pharmaceutical industry

connected, surrounded by and dependent on outer *layers*, where creativity is more or less represented as a social construction. The core is well exemplified by the processes that lead to concepts and ideas that may lead to the discovery of a new chemical entity (NCE), which, in a successful case, later becomes a candidate drug (CD). The novelty and usefulness of the NCE and the CD is always scientifically proven, using valid and commonly accepted evaluation methods. Thus the NCEs and CDs represent some kind of *biological and medical truth*, or what Whitehead (1925) calls 'irreducible and stubborn facts', which fulfill the criteria of novelty (appearing for the first time) and usefulness (showing the benevolent effect of treating a disease).

These factors are determined in an objective way using reliable methods. For example, an approved substance patent signifies the creative output. Thus, the objectivity of an NCE is also valid outside the organization or the industry, and for the entire scientific community at large. But the creation of an NCE cannot stand on its own, or in Heidegger's (1977) formulation: 'science cannot speak for itself'. This means that the NCE is not a drug product or medicine – it must be developed further to become a finished, approved product on the market, in a sense, the final innovation. Thus, *inner-core creativity* demands a multiplicity of creativity at different stages, and in various ways, during the entire R&D process. This is represented by the layers, which can represent creativity as a social construction in different ways.

Creativity in the layer(s) is much more pragmatically driven to create ideas and solve problems needed to further develop and document the

NCE to be realized as a finished product. The socially constructed part can, for example, be the constant intervention of what Csikszentmihalyi (1999) calls the *field* (i.e., those individuals give to prerogative to judge what is creative and what is not, in organizations generally managers and experts). For example, management intervenes in the nomination of an NCE to become a CD, *deciding what is or is not creative*, based on different criteria (e.g., market potential and fit in relevant therapeutic areas). Other examples of this kind are new ideas for developing methods, devices, technologies, services, processes, and practices. In contrast to the inner core, the layer can be seen as more pragmatically driven. For example, during a clinical study, creativity may involve finding new ways of designing clinical studies that involve new sets of parameters and variables (while following commonly accepted guidelines and standard operating procedures), which might result in the discovery of a new indication.

Yet another example is that many drug molecules often display properties or characteristics that make it complicated to acquire an optimal therapy and administration. Thus, layer creativity in development can be ideas that lead to finding the appropriate and optimal pharmaceutical formulations of the drug, together with relevant analytical technology to document the finished product. A third example could be creative solutions that lead to new processes and practices for scientific information collection and assessment, e.g., large multinational clinical trials, which allow researchers to interpret data faster and, in the end, enable the organization to reach the market faster. A final example of creative output from the outer layer can be found in ideas for branding the product (e.g., logos, marketing strategy, pricing, and information packages for physicians), which, in the marketing phase, successfully promote the product. So the core and the layer(s) of creativity in new drug development are interconnected, yet separable. Important aspects of the model are the *translation interfaces*. These interfaces are important – particularly between the core and layer for supporting creative action by translating and diffusing knowledge and information and for stimulating connectivity between different projects. For example, *Scientific Champions* (i.e., leading experts in therapeutic areas) illustrate this role. The political actions influenced projects by providing scientific contributions, as well as the new ideas and commitment that are a function of an extraordinary personality.

The creativity process (e.g., recognition, selection, and evaluation) is more homogeneous in the core – in contrast to the layer. What is or is not creative in the core is more easily agreed upon because arguments are less complicated; they are based on hard, uncontested scientific facts and scientific practices, embodied in laboratory work and often supported by external actors and communities. Creative action in the layer is more complex and heterogeneous by nature. The layer represents projects in their later phases, involving more resources and costs in comparison to the core. What is or is not creative in the layer is more contested, more socially constructed, and involves divergent influences (e.g., the market, safety, and regulatory bodies). So an important implication of the model is that the creativity will become more problematic, contested, influenced by political struggles and internal competition when moving from the core out to the layers (e.g., during later development phases), from the scientific procedures and practices to business objectives and financial interests.

Science-based creativity in the pharmaceutical industry is similar to what Popper (1963) terms the 'problem of demarcation of science'. Popper developed an objectivist theory of science that became known as falsificationism. Popper intended to show the distinction between science and non-science by means of its falsifiability, or refutability, or testability. The notion of falsification formulated by Popper, held that to come closer to the truth, scientists should invent bold hypotheses that are testable and discarded as soon as counter-evidence is discovered, thus providing convincing rational answers for what science and knowledge really are (Popper, 1963).

Lakatos (1970) developed an alternative view of Popper's thinking that he calls 'sophisticated falsificationism' in contrast to what he deemed as Popper's 'naive falsificationism'. According to Lakatos, a hard core (i.e., central theory) characterizes all scientific programmes. The hard-core theory is protected from refutation by the negative heuristic which is the instruction: as far as possible, fit theory to results by introducing amendments in the auxiliary hypotheses (and/or initial conditions). His main point is that theories of a certain kind, the kind that is the cores of research programmes, are not falsified in practice. They can be cumulatively disconfirmed over a time period, but they cannott be decisively knocked out by a single crucial experiment.[21] This aspect is also reflected in what is called Duhem's hypothesis: refutations are centered at the technology employed rather than the theory itself (Hacking, 1983).[22]

---

[21] Lakatos (1970: 133) gives an example in Newton's three laws of dynamics and the law of gravity. He writes: 'This core is irrefutable by the methodological decision of its protagonist: anomalies must lead to changes only in the protective belt of auxiliary, observational theories supporting these anomalies.'

[22] After Pierre Duhem (1861–1916), French physicist, philosopher and mathematician.

The main point of this discussion is to show that the demarcation between objective and materialist and socially constructed creativity in the pharmaceutical industry is complicated and distinct from the view of creative output in other industries.

For example, in the automotive industry (e.g., concept cars) or the telecom industry (e.g., new designs or functions of mobile phones) organizational creativity is not merely socially constructed. But the *core* creativity in new drug development is based on strict hard science and technical-instrumental rationality, although there are layers of more or less socially constructed creativity that are crucial to take the finished product to the market – to make the final innovation.

Conclusions

In this chapter, three different epistemological aspects of creativity have been examined – namely, the position of the subject in creative work, the difference between discrete and connectivistic perspectives on creativity, and the distinction between social constructivist and objectivist and material definitions of creativity. These three perspectives on creativity draw upon ontological and epistemological thinking and seek to turn back to the theoretical and philosophical underpinnings of the notion of creativity. Without a proper critique of creativity *qua* epistemologically founded construct, it remains a fragile construct. Needless to say, a certain theoretical construct, a signifier used in a language game, can be associated with virtually any epistemological position and as a consequence, there may be a broad range of perspectives on what creativity is and how it functions in practice. In the view pursued in this book, creativity is, as Whitehead suggests, a form of actualization, a series of connection and associations within a realm of practice, for instance, new drug development in the pharmaceutical industry, based upon the alignment of humans, technology, theoretical frameworks, non-humans (e.g., laboratory animals) and a number of additional resources. Creativity is therefore distributed, dispersed, non-linear, and at times even chaotic. From a managerial perspective, the management of creativity is therefore not a trivial matter. However, being capable of maintaining an image of creativity that recognizes all these heterogeneities and complexities enables a more adequate management practice. Pursuing simple models of reality does not of necessity promote better management practice.

4
Exploring Creativity in Organizations: Methodological Concerns

Introduction

In this chapter, the use of different methodological aspects, combined with the three influential theoretical models, will be critically examined in terms of relevance for the study of organizational creativity and a management study, which in this case means support, reinforce and increase understanding of the creative activities in organizations. The argument is that practically all current work on creativity is based on methodologies that either are psychometric in nature or were developed in response to perceived weaknesses of creativity measurement (Plucker and Renzulli, 1999). The main bulk of creativity literature is either conceptual or in the realm of quantitative methodology. In addition, the majority of creativity studies have generally focused on only one level of analysis at a time (e.g., Taggar, 2002). There is an ongoing debate about the appropriate level of analysis in studying creativity in organizations. Traditionally, creativity research has concentrated on the small group (or independent project) as the focal level of analysis. With some exceptions (e.g., Glynn, 1996; Woodman *et al.*, 1993), little has been done to extend research beyond the level of the small project (Drazin *et al.*, 1999).

There is a large gap between the prevalent use of methods (primarily quantitative methodology) to examine creativity in organizations, combined with the need to acquire generalizability, and the applicability and sense making to other organizations. The strong reductionist and functionalist traditions in creativity research (inherited from the psychometric tradition) have been concerned with issues such as validity and reliability, and dealing with an all-encompassing definition of creativity. Another dilemma in creativity research methodology surrounds the confusion of the construct itself. This is exemplified by the subject-centred view of creativity vis-à-vis the more distributed image of creativity in organizations. This is exemplified by Wehner *et al.* (1991: 270) who describe the situation pertaining to creativity research methodologies in terms of the fable of the blind men and the elephant: 'we touch different parts of the same beast and derive distorted pictures of the whole from what we know: the elephant is like a snake, says one who only hold its tail; the elephant is like a wall, says the one who touches its flanks'. Commenting on this metaphor, Plucker and Renzulli (1999: 50) write: 'The challenge to creativity researchers, especially employing psychometric methods is to distinguish between the elephants (various conceptualizations of creativity) and the domestic pets (barely relevant constructs and extraneous factors influencing creativity productivity.' As a consequence, when it comes to investigating the complex and collaborative nature of creativity in organizations, much research and literature has been caught in a methodological trap which has resulted in myopic and static perspectives, leaving us with few opportunities to more critically renew methods in order to create an increased understanding of creative activities in organizations. One side of this aspect is exemplified by Plucker and Runco (1998: 37), who argue that when people engage in creative activity 'their thoughts and actions are guided by personal definitions of creativity and beliefs about how to foster and evaluate creativity that may be very different from the theories developed by creativity experts'.

One of the fundamental problems in investigating organizational creativity is to define and measure the dependent variable. This construct definition is also consequential for theory building (Sternberg and Lubart, 1999). In general, scholars have defined creativity as an important outcome from a system. It is seen as an independent variable and treated as one of the factors to be manipulated in order to improve the outcome of this approach. This functionalist and reductionist view has dominated creativity and innovation literature (Rickards, 1991; Drazin, 1990). Furthermore, a large part of the organizational creativity literature is essentially theoretical or conceptual (e.g., Woodman *et al.*, 1993; Drazin *et al.*, 1999; Glynn, 1996) and empirical studies aiming to understand organizational creativity are scarce (Ford, 1995a).

This chapter is structured as follows: First, we offer an overview of traditional methods of single perspective of the subject-centred view of creativity. Secondly, three evolving influential system theories representing confluent perspectives in organizational creativity research are discussed. Thirdly, an alternative methodological approach is presented, here denoted as the *multiparadigmatic approach*. This approach is not limited to mixing methods or theories; it also acknowledges the importance of including action research and collaborative management research perspectives in organizational creativity research. Then follows an example how this approach can be used to organizational creativity research relevant to pharmaceutical R&D. Finally, some practical and theoretical implications are outlined.

Overview of methods in creativity research

Single perspective methods

Creativity research has been previously dominated by quantitative methodology used primarily to assess not only the *creative person* and *process*, but also *the product and the place* (i.e., environment) (e.g., Plucker and Renzulli, 1999; Amabile, 1988; Ekvall, 1996). Psychometrics is the umbrella term for methods of assessing personality traits and mental representations and processes underlying creative thought such as divergent thinking (e.g., fluency, flexibility, originality, and elaboration) (Eysenck, 1996; Sternberg and Lubart, 1999). This line of research includes longitudinal studies and uses different instruments,[23] mostly in non-working areas (Plucker and Renzulli, 1999). Another method is historimetry, which is the application of quantitative methods to archived data about notable figures from the past (Simonton, 1984). Amabile's (1982) *consensual assessment technique* (CAT) is one of the most-cited methods for assessing creative products. An important reason for this is that when researchers use different definitions of creativity and thus different criteria for assessing creativity, it is difficult to compare their research findings. Amabile (1982) suggested a method based on an operational definition of creativity, which implies that a 'product or idea is creative to the extent that expert observers agree it is creative' (Amabile, 1988: 14) wherein no criteria are given and the judges evaluate independently. In assessing the dimension of the creative place or work climate in organizations, there are many empirical studies undertaken using instruments – for example, Amabile's (1999) KEYS instrument (used for assessing the creative climate) which contains a 78-item inventory that covers different environmental scales of obstacles and stimulants for creativity. Another example is Ekvall's

---

[23] Examples include: the test of creative thinking (TTCT) (Torrance, 1974) or the structure of intelligence (SOI) test for divergent thinking (Guilford, 1967).

(1996) 50-item instrument called the creative climate questionnaire (CCQ). The approach most distinct from the psychometric approach is the narrative or case study approach (Gedo and Gedo, 1992; Gruber and Davies, 1988) in which researchers construct case studies. According to Plucker and Renzulli (1999: 38), this methodology is still in its infancy compared to the others (i.e., psychometrics).

Confluence perspectives

There is a strong belief in some communities that to increase the understanding of creativity, a multidisciplinary approach is required (e.g., Gardner, 1993; Gruber, 1989). Many recent methods for studying creativity hypothesize that multiple components must converge in order for creativity to occur. Regarding the confluence of components, creativity involves three important components to consider: *knowledge, motivation*, and *environment* (Sternberg and Lubart, 1999).

Determinants and methods for organizational creativity

The literature draws attention to five major organizational factors that influence creativity in the work environment: organizational climate, organizational culture, leadership style, resources and skills, and the structure and systems of an organization (e.g., Amabile, 1996; Ekvall, 1996; Locke and Kirkpatrick, 1995). Scholars argue that these factors create conditions that enhance creativity at team and individual levels (Woodman *et al.*, 1993). Recently, several papers attempted to portray theories of organizational creativity (e.g., Kazanjian *et al.*, 2000; Ford and Gioia, 2000), but empirical studies applying these models in practice are scarce (Ford, 1995). Scott and Bruce (1994), who employed data (questionnaires) from the R&D organization of a large corporation, report one such approach. In their analysis, using structural equation modelling and path analysis, they view creativity as an outcome of four interacting systems: individual, leader, work group, and climate for innovation.

Influential systems theories of organizational creativity

The systems perspective of creativity in organizations

Systems theory is integral to an understanding of the system's context. Generally, this view maintains that the whole is more than the sum of its parts. Most systems theorists stress that everything is an open system and that the interaction with other systems in the environment influences the organizational development (Capra, 1996). These theorists (e.g., Bertalanffy, 1968) contend that the interaction of an individual component within a system allows us to reach a deeper understanding of systems. General systems theory, originally developed by the biologist Ludwig von Bertalanffy, tries to explain different observable phenomena as 'wholeness'. In contrast to many other theories in various sciences which try to explain observable phenomena by reducing them to interplay of reduced and elementary units (often treated independently), systems theory attempts to take into account the interactions in a system. Wholeness in this context means very broadly, *'problems of organization and phenomenas are not resolvable into local events, dynamic interactions in the difference of behavior, manifest when isolated or in a higher configuration'* (Bertalanffy, 1968: 37). In short, systems of different sorts are not understandable when investigating their respective parts in isolation.

The DIFI model of creativity developed by Csikszentmihalyi (1988) is a theoretical framework that defines creativity as being dependent upon persons, processes, products, and places. This theoretical framework is now widely accepted as useful in understanding organizational creativity. The basic argument in this view is that creativity should be defined as a socially constructed label that is used to describe actions embedded in particular contexts (Ford and Gioia 2000). According to Csikszentmihalyi's DIFI model, creativity must be defined with respect to a system that includes individual, social and cultural factors that influence the creative process and help to bring about a creative outcome. This systems approach describes three interrelated subsystems: the *domain*, the *field*, and the *individual* (see Figure 4.1). One important implication of the model, according to Csikszentmihalyi, is that the level of creativity in a given place at a given time is not solely dependent on the amount of individual creativity. It depends just as much on how well suited respective domains and fields are to the recognition and diffusion of novel ideas.

The first subsystem, the *domain*, consists of the symbolic system of rules and procedures that define a system with its own set of symbolic elements, knowledge, rules, and notations. One important general characteristic of the domain is that every domain has its own internal logic and its characteristic pattern of development. Those who operate within it must respond to this logic. For instance, the scientific discipline of biochemistry can be seen as a specific domain that consists of various axioms, practices, and rules.

When applying this concept, almost any human activity can be seen and framed into different domains and subdomains of knowledge and

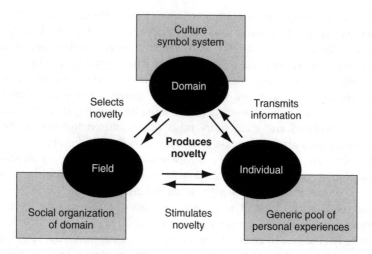

Figure 4.1   Csikszentmihalyi's (1999) domain individual field interaction (DIFI) model

activity – from football teams to scientific disciplines and to corporations. Using this definition, a domain can be exemplified at different levels in an organization, including different functions and skills that represent a specific body of knowledge, language, and customary practices. The domains in organizations are presented as 'given knowledge', the basic factors of the profession which in practice often involve creativity in the sense that creativity is necessary to identify areas that can be intelligently and cost-effectively improved (Csikszentmihalyi and Sawyer, 1995). The second subsystem, the *field*, includes the gatekeepers, managers, experts, or stakeholders who personify and affect the structure of a domain and who are entitled to select a novel idea, service or product for consideration. The field within a domain also has the power to change it. Csikszentmihalyi talks of the field as including 'all individuals who act as gatekeepers or managers to the domain'. The gatekeepers' function is to decide whether a new idea of product should be a legitimate part of the domain. Gatekeepers are all those people whose roles in a creative ecosystem give them the power to decide whether or not particular creative acts or products are placed into channels of transmission or creative outlets by which they can become visible to relevant audiences (Harrington, 1999).

Gatekeepers in the domain of mathematics, for example, are those distinguished professors and journal editors who decide whether or not a new contribution to the domain is to be published. So the field consists of experts and authorities who are responsible for passing judgements on performance in the domain. This responsibility creates competition between the individual and the group – to convince the field that the person or group has a valuable innovation. In many organizations, different management teams play this role. The last subsystem, the *individuals*, is the person or the group that produces the novelty. These three subsystems jointly bring about the occurrence of a creative act. The primary role of the person is to introduce variations within a field. The gatekeepers or managers, who comprise and represent the domain, select from among these variations (novel acts). So, according to the systems approach, creativity always occurs within specific configurations of knowledge, and there can never be any creativity *as such*; creativity is always creativity with others.

Some interesting implications of the model can be noted: (i) the model emphasizes a crucial step in the creative process: innovation can only be secured when the actual idea or novelty is selected and accepted by the appropriate field or management and implemented into a relevant domain; and (ii) the model overcomes the dichotomy of over-socialization and undersocialization through aligning the system's view (the domain) with the actor-perspective of the gatekeepers and the creative individuals. Another important aspect of the theory is the fact that creativity cannot be separated from its recognition. Csikszentmihalyi (1996) illustrates this aspect with an example from the domain of music. The conventional explanation is that J.S. Bach was a creative composer. But his music was actually dismissed as old-fashioned for several generations until it was rediscovered by Felix Mendelssohn, a representative of the field during the mid-nineteenth century. This rediscovery resulted in Bach's full recognition as a creative composer. This example implies that we are constantly reassessing the past *and* creativity.

Finally, this theoretical approach: (i) provides opportunities for a better understanding of new product development activities, such as those in the pharmaceutical R&D process, including the discovery process; and (ii) views the development stages of pharmaceutical research as creative processes. The model does not restrict creativity to artistic expression, but claims that all domains enable creative extensions of what can be done. Furthermore, the model also emphasizes the importance of management's role in the creative process. This notion that creativity in an organizational context – as an interaction between individuals within a domain and gatekeepers (managers, peers or experts) who reject or retain creative action for future or further implementation – should not be restricted to a single domain (e.g., a department, a function, or a scientific discipline). The result is that creative actions in organizations often face overlapping, multiple domains rather than single domains (Ford and Gioia, 2000; Ford, 1995b). Thus, this view of creativity in organizations includes the importance of the interaction of many domains in the organization in which different informal social networks play an important role (Bras, 1995).

To conclude, using the DIFI model, and the metaphor of the blind men and the elephant, one may say that the men now actually grasp the idea that the elephant and other pets are living together in an environment; it's not only the elephant that is important. However, the image is blurry and fragmented, but one thing is for sure – someone is riding the elephant.

The interactionist model of organizational creativity

This perspective is grounded in interactional psychology (Bowers, 1973). Generally speaking, the interactionist approach to personality is based on the assumption that an adequate description of an individual's behaviour can only be possible if the context in which it occurs is taken into account. The interaction between personality and situational variables is the basis of the interactionist approach. Bowers (1973: 324) characterized the interactionist approach best when he wrote, 'The Skinnerian legacy of studying one organism at the time clearly has its virtues. However, employing this strategy makes it virtually impossible to see how different situations affect different individuals differently; the very possibility to for an interaction term disappears.' Researchers have examined the theoretical underpinnings of socialization – both in content and process – and empirical studies have moved this work forward, but have examined them either primarily from the individual's or the organization's perspective. The interactionist perspective aims to integrate these two areas by examining how individuals' attempts at self-socialization work in tandem with the organization's attempts at socialization to influence socialization outcomes (Griffin *et al.*, 2000). Some psychologists consider interactionism as a conditional version of trait theory: they limit themselves to the study of the mechanistic interaction between people and situations – for example, by modelling statistical interaction of ANOVA (Rouxel, 2001). The basic tenet of the interactionist approach is that human behaviour must be understood as a product of person and situation.

Putting this perspective into creativity research implies that individual differences in creativity may be partly understood in terms of individual characteristics, such as cognitive style, cognitive ability, personality, and motivation. But situational and contextual factors are also important (Woodman and Schoenfeldt, 1990). This means applying an interactionist model on creative behaviour at the individual level. Woodman, Sawyer, and Griffin (1993) were the first researchers to take a clear step forward in trying to define the concept of organizational creativity and to integrate previous research into more of a systems approach. They extended this interactionist perspective of creativity into the organizational arena, where creativity is viewed as a complex outcome of person and situation. The situation can be analyzed in terms of the social and contextual influences that either facilitate or inhibit creative accomplishment. They propose an interactionist model of creativity at individual, group, and organizational levels within an environmental context. In this model, individual creativity is a subset of team creativity, which subsequently is a subset of organizational creativity. This model proposes a recursivness, or a succession of elements that relate to each other within an organization, which can be seen as nested subsystems. The model adopts an interactionist approach, which intends to retain a linkage between the four subsystems of creativity: process, products, person, and place (Schoenfeldt and Jansen, 1997).

According to the model, individual creativity behaviour is a function of antecedent conditions, cognitive styles and abilities, personality, motivational factors, and knowledge, which implies that individual creativity contributes through group-level creativity to the organizational level. Group-level creativity behaviour accounts for group characteristics, group processes, and social information processes. The model particularly emphasizes the contextual influences in the interface between individual and groups. The model also integrates the environment or organizational climate as an important factor for creative outcome. The conceptual model takes a step towards understanding organizational creativity as a combination of the creative process, creative product, creative person, and creative situation – and the way in which these components interact. The situation is then characterized in terms of the conceptual and social influences (that either facilitate or inhibit creative accomplishment). Here, Woodman also emphasizes that the creative behaviour of organizational members is a complex person–situation interaction influenced by events of the past and prominent aspects of the current situation. Or, as Woodman and Schoenfeldt put it: 'From an interactionist position there is always something more to understanding behavior than just describing the observed behavior *per se*. This "something more" has to do with the essence of the organism and its behavioral potentiality' (Woodman and Schoenfeldt, 1990: 296).

The componential theory of organizational creativity and innovation

The componential theory suggests that that the area of overlap between the elements conveys 'the area of highest creativity for individuals and highest innovation for organizations' (Amabile, 1988: 157). Amabile also recognizes that environmental models can serve either to promote or inhibit creativity in organizations (Amabile, 1999a; 1997). According to Amabile, action must be taken by management to nurture innovation and allocate resources for its development and implementation. She defines organizational innovation as 'the successful implementation of creative ideas within an organization' (Amabile, 1988: 126). The theory includes three major components of individual- (or group-level) creativity; each component is necessary for creativity in any given domain. The theory uses three interlocking circles to represent each of the three components of creativity domain-relevant expertise, creativity-relevant skills and processes, and intrinsic motivation (task motivation).

According to Amabile, expertise or domain knowledge is the foundation of all creative work and is seen as a set of cognitive pathways combined with memory for factual knowledge and technical skills in the target domain. Creativity skills or creativity-relevant processes rely somewhat on personal characteristics, such as tolerance for ambiguity, self-discipline, orientation towards risk-taking, ability to explore new cognitive pathways, and working style. The last element in the model – task motivation or intrinsic motivation – is seen as a fundamental driving force for creative action in organizations and requires a more flexible view of attention and support. This notion derives from the *intrinsic motivation principle* of creativity, which suggests that people will be at their most creative when they are primarily intrinsically motivated by the interest, enjoyment, satisfaction and challenge of the work itself (Amabile and Conti, 1999; Amabile *et al.*, 1996). While extrinsic motivation relates to factors in work that are driven by the desire to attain some goal outside the specific work tasks, such as achieving a promised reward or position or meeting a deadline. Research has shown that high intrinsic motivation and relatively low extrinsic motivation induce creative individuals to be more independent of the domain of knowledge and less susceptible to pressure to conform (Amabile, 1999b, 1997). Intrinsic motivation is also offered as an explanation for why creative people show great involvement and energy in their tasks (Deci and Ryan, 1985). Another way to consider the importance of motivation in organizational creativity lies in the strong tradition in the organization of reliable support of creative techniques and in close focus on domain-relevant knowledge and cognitive abilities – abilities that may be irrelevant without motivation skills. From the model Amabile proposes, four criteria for models of organizational innovation: (1) the entire process of individual creativity must be incorporated; (2) all aspects of organizational influencing innovation should be considered; (3) the phases in the organizational innovation process should be profiled; and (4) the influence of organizational creativity on individual creativity should be described (Williams and Yang, 1999).

Critique of the theoretical models and methodology

The previous three systems theories on creativity, which are somewhat interconnected and exert considerable influence on one another, have several advantages for the study of organizational creativity. The DIFI model moves from a focus on the individual to a systems perspective (e.g., organizations), which includes the effects that cultural context and role management (i.e., the field) have on the creative process. The interactionist model views creativity as a complex outcome of the person and the situation, but the model retains a link to previous research (person, product, process, and place) in which new perspectives and methods for organizational creativity may be found (Schoenfeldt and Jansen, 1997). Within a systems-based view, creativity can still be seen as an individualized phenomenon (Sternberg and Lubart, 1999), but the creative process is perceived as occurring within the context of a particular environment rather than a vacuum (Williams and Yang, 1999). Amabile's componential theory of organizational creativity encompasses the importance of domain-specific skills and the role of intrinsic motivation, which also provides a useful way to conceptualize the importance of the social environment in creativity, that can support or undermine the intrinsic motivation to create. However, this Componential Theory and the tests she developed to measure the component processes still missed a lot.

In different ways, all three models address the managerial aspect of organizational creativity. All models reject the notion that the all-encompassing definition of creativity is not well suited for organizational creativity: generalizability should be treated within a narrowed range within the specific organization and must be incorporated in the organization's mission and market, emphasizing dimensions of the actual work and of value (Sundgren and Styhre, 2003). Some benefits and limitations of these models are presented in Table 4.1.

Table 4.1   Some benefits and limitations on confluence perspective model of organizational creativity

Model	Benefits	Limitations	The blind men & elephant metaphor
DIFI model (Csikszentmihalyi, 1988)	Moves from a focus on the individual to organizations, and link to managerial influence.	Avoiding the complex and undesired effects of organizational creativity	The men grasp the idea that the elephant and other pets are living together in an environment and that someone is riding the elephant.
The interactionist model (Woodman, 1993)	Admit complexity and emphasize the contextual influences in the interface between individual and groups.	To mechanically retaining previous research (e.g. person, product, etc.). Viewing organizational creativity as a causal aggregation (i.e. person to company level)	The men have recognized that the elephant have legs, ears and tusks. They agree that these are important, and, that there are probably some more to discover.
The componential model (Amabile, 1999b, 1997)	Acknowledge domain-specific skills and the role of intrinsic motivation.	Narrow and limited perspectives.	The men understand some clues about the behaviour of the elephant; what actually drives the elephant to eat and move.

These models can be seen as versions of the systems approach in which the wider system interacts with an environment, and component subsystems may be found. But these models fail to address several important issues that extend the understanding of organizational creativity. Such issues can link varied perspectives on measuring and assessing data: Woodman's model, for example, is a kind of reductionism model of creativity in an organization. It presupposes creativity and causality between components. As Rickards and De Cock (1999) point out: 'assessing team creativity is not a matter of aggregating individual creativity, nor can the performance of a set of teams be aggregated to assess an organization's performance'. Other issues are informal networking, information sharing, and management practice dimensions. Although these models offer a good theoretical platform for organizational creativity, further adaptation, is needed for a particular organization (which most probably may include the implementation of new constructs). Furthermore, they are not fitted only to the use of quantitative methodology. As Csikszentmihalyi (1994: 154–5) writes:

> [I]t is unlikely that creativity research will ever become an entirely independent symbols system with its own special theoretical constructs, methods, and procedures. Instead, it is more likely to become an interdisciplinary domain in which humanist and social and biological scientists retain their own conceptual tools and approaches but find a way of integrating them to study processes that do not admit one-dimensional explanations.

A multiparadigmatic approach to creativity research

A multiparadigmatic approach does not necessarily accept that all definitions of creativity are equal, but it does open avenues for a new kind of progress that allows various designs and research approaches to be used – depending upon the type of organization. Rickards and De Cock (1999) argue that creativity is inherently a social concept regardless of whether the focus is individual, organizational, or societal. Every researcher acts from within a web of social relations that connect different influences. The basic argument for using a multiparadigmatic approach is to enrich the study of creativity in organizations by exploring and bridging different paradigms and thus generating new theories that may be more suitable for the study of organizational creativity in a given organization.

Ever since Kuhn (1962), the term *paradigm* and the concept *creativity* have been used in a variety of ways. Rickards and De Cock (1999: 240) write: 'We see a paradigm as a set of internally consistent and simplifying heuristics that inform individual and social action'. This is consistent with Burrell and Morgan's (1979) understanding of paradigms as ideal types of opposing meta-theoretical assumptions that can be treated as worldview or reality assumptions. One useful way of positioning past and present creativity research is through Rickards and De Cock's (1999) paradigmatic analysis of creativity research. This is based on Burrell and Morgan's assumption that mainstream thinking in social science can be studied by mapping any coherent theory, or body of enquiry, along two distinct dimensions: subjective/objective and regulatory/revolutionary (Parker, 2000). Rickards and De Cock attempt to locate creativity research, and its different main contributors, within these four paradigms (see Figure 4.2).

Rickards and De Cock's (1999) paradigmatic mapping helps to explain confusions in the literature on creativity. There is confusion around the definition of creativity. For example, the functionalist view may 'define' creativity as the process, whereas, for the interpretive perspective, it is more a matter of personal reframing. Terms such as *truth* and *validity* will also have different meanings. Thus, a multiparadigmatic approach may be a way of assessing the relative merits of each paradigm. Or, as Rickards and De Cock (1999: 249) point out: '[t]he multiparadigmatic approach will offer considerable rewards for understanding the nature and stimulation of creativity at individual, team, and organizational levels'.

The functionalist paradigm (objective/regulatory)

For historical reasons, this paradigm captures the most widely accepted theories. It is in line with a positivistic tradition emphasizing that

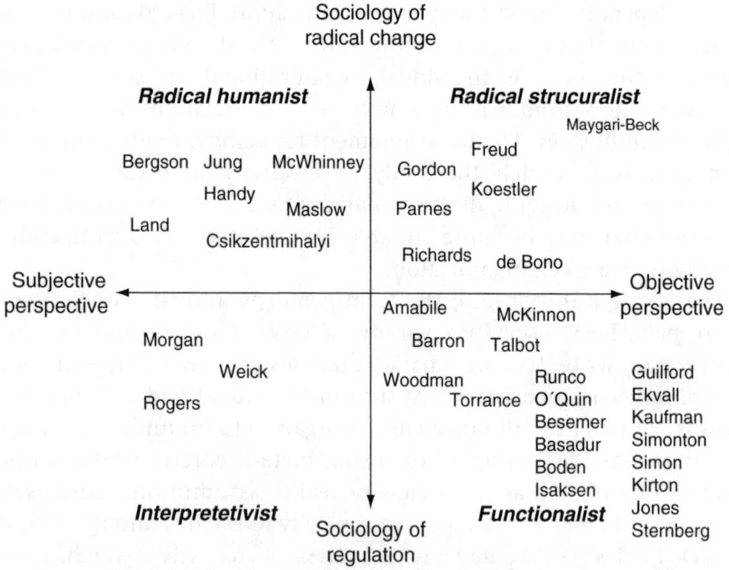

Figure 4.2 Four quadrants for meta-paradigmatic analysis and overview of creativity research mapped onto the Burrell and Morgan (1979) matrix (adapted from Rickards and De Cock, 1999)

measuring, but avoiding interacting with, the system is the best method of gaining generalizable knowledge. Attention is directed to linking truth to the confirmation of predicted results and empirical regularities. According to Rickards and De Cock (1999), this paradigm also involves: (i) pursuit of the correct definition of creativity and (ii) conducting large surveys, whose results are generated using statistical analysis. Theories within this quadrant include all four perspectives: person, process, product and place of creativity research. The primary objective is not to pass judgement on or transform the world but to explain phenomena.

Practical example 1: Drivers of organizational creativity: a path model of creative climate in pharmaceutical R&D (Sundgren *et al.*, 2005a). This study was generated from an opportunity within Medical Informatics[24] in AstraZeneca, where the main authors have held a professional position. The study is based on quantitative data from an original questionnaire covering a conceptual framework for organizational creativity in pharmaceutical R&D in which information sharing,[25] networking, learning culture, and intrinsic motivation were hypothesized to affect the perceived creative climate in two global organizations in Development R&D in AstraZeneca. The theoretical framework is based on Woodman *et al.*'s (1993) interpretation of research findings within the interactional perspective of organizational creativity. Woodman *et al.* (1993) proposed three propositions and 12 hypotheses to guide further research on organizational creativity. Four subsets of these hypotheses were investigated in the study, which includes the role of informal networks, the need for a culture of learning in an organization, and the role of motivation. The study uses confirmatory factor analysis (CFA) to investigate whether items measure the hypothesized dimensions (representing several research streams in organizational creativity research) and whether the hypothesized dimensions are empirically differentiable. Structural equation modelling (SEM) was used to (1) evaluate the causal and

---

[24] Medical Informatics was a strategic change initiative. Its long-term objective is to develop new business models and proposals within the global development organization of AstraZeneca. The purpose is to (1) enhance efficiency and to nurture organizational creativity by improving clinical researchers' capabilities for exploiting and exploring scientific information globally; and (2) support informal networks.

[25] Information in this study refers strictly to different types of scientific information that in some form are relevant to AstraZeneca's research projects.

correlative links between theoretical variables; and (2) develop the final path model.

The interpretive paradigm (subjective/radical)

This perspective replaces the objective truth of the functionalist paradigm with symbolic truths that are revealed in stories, narratives, and social transactions. Attention is directed away *from* measurements based on the outer world of physical realities *towards* the inner world of feelings, needs, concerns, and values. This paradigm includes research positions, which state that the process of studying creativity may be more fruitful by becoming a part of reality, when the role of the researcher is that of interpreter of the emerging story. A typical example of this perspective is the emphasis of social psychology and action research methodology. The objective is to explain – with an emphasis on facilitating a process of reconciliation of differences.

 Practical example 2: Creativity – a volatile key of success: creativity in new drug development (Sundgren and Styhre, 2003). This study explored projects in former AstraZeneca (ICI[26] and Astra[27]) from an organizational creativity perspective by using qualitative methodology (interviews). These two companies were very successful in new product development, with several blockbuster drugs that were developed between 1975 and 1985. The study investigates new product development (including discovery and development) activities during the period; it focuses on seven successful projects within the two companies. The study adopted a systems theory perspective; here, creativity can be seen as an emergent property within a sociocultural context that is shaped by multiple forces, including – but not limited to – contributions of the individual. One important argument for using this approach is that it aims to describe relationships within a system that is important for creative action – rather than describing a single cause of the origin of creativity. The study is based on recent interviews with many of the most influential researchers, project leaders and line managers, who held positions in the seven projects during the period under consideration. Central findings

---

[26] Refers to ICI's (Imperial Chemical Industries) pharmaceutical division, located in Alderley, UK, that in 1993 became the R&D organization in Zeneca and is today one.

[27] Referring to AB Hässle, a subsidiary R&D Company of Astra, located in Mölndal, Sweden.

in the study are that if creativity is to be managed as an organizational resource, there are at least nine factors that must be considered. These factors range from practicalities to issues of project culture and human faculties, such as curiosity and intrinsic motivation.

The radical structuralist paradigm (objective/radical)

The radical structuralist believes that with an objective view, radical change can be achieved and discovered through theoretical principles that replace current and traditional orthodoxy. The notion is that creativity research is too fragmented and should instead strive to develop new principles of triggering physiological states and new problem-solving systems. The aim is to replace structures and behaviours and to support innovation and change.

Practical example 3: Dialogue-based evaluation as a creative climate indicator: evidence from the pharma industry (Sundgren *et al.*, 2005b). This study examined how different forms of performance evaluation affect aspects of the creative climate in AstraZeneca R&D using quantitative methodology (Multivariate analysis, ANOVA). Data used for the analysis are from a recent global employee questionnaire survey at AstraZeneca.[28] The study was based on data exclusively from the R&D organization. The responses in this study came from 5,333 employees, including the development and discovery organizations within five R&D sites (three in Sweden and two in the UK) and thus represent a majority of the R&D sites and more than 50 per cent of the company's global R&D organization (53 per cent response rate). Thirty-one items were extracted from the global survey study, based on their relevance to motivation (intrinsic and extrinsic), value-focused thinking, control-based evaluation, dialogue-based evaluation, and organizational creativity. The study focuses on: (i) the impact that management's evaluation of employees – either dialogue-based or control-based – has on the type of motivation (intrinsic or extrinsic) that drives employees or their style of thinking – value-focused thinking; and on (ii) their attitudes towards organizational creativity. The theoretical framework is based on the

---

[28] FOCUS I survey, conducted in 2000, addressed the entire AstraZeneca organization, including marketing, production, and research companies and had 138 items that covered a wide range of organizational issues, such as organizational function, education background, opinions about daily work life, communication, management, and external competitors. More than 38,000 employees were invited to respond to the survey at AstraZeneca.

hypothesis that situational factors (Shalley and Perry-Smith, 2001; Deci and Ryan, 1980, 1985; Ryan, 1982) can affect behaviour related to creativity in two ways: one is *control-based*, and the other is *dialogue-based*. The central finding in the study is that dialogue-based evaluation is a better indicator for intrinsic motivation and organizational creativity – compared to control-based evaluation. The study argues that dialogue-based evaluation can bridge and reduce discrepancies between the *assumed* and *politically correct* culture versus the *enacted* and *true culture* and thus become one way to manage creativity in an age of management control.

The radical humanist paradigm (subjective/radical)

This paradigm places less trust in traditional notions of social science, and instead seeks a more subjective and non-traditional approach to replace them. Attention is focused on the need for creativity to transform organizations in times of organizational turbulence. Creativity is highly desirable and is essential to allow escape from the inefficiencies of traditional cultures, structures and practices. Thus, the objective is to change.

Practical example 4: Intuition and pharmaceutical research: the case of AstraZeneca (Sundgren and Styhre, 2004). This study explored the role of intuition and its implications for organizational creativity within pharmaceutical R&D by using qualitative methodology (i.e. interviews). The main reason for conducting this study was that previous studies of scientific organizations show that scientific work is never as linear, homogeneous, and one-dimensional as might initially be imagined. Instead, controversies, alternative explanations, empirical inconsistencies, and local interpretations always characterize production of 'scientific facts'. In short, a certain degree of heterogeneity exists within scientific knowledge. The hypothesis in the study is that intuition constitutes an important ability to apply scientific knowledge and to see the consequences of various experiments before formal proof is acquired. The theoretical framework in the study applies the philosophy of French philosopher Henri Bergson, in which intuition plays a significant role, i.e., knowledge is separated through use of ready-made concepts, and intuition is the ability to think between these concepts – to think between the known and the abstract. Intuition can be described by using a popular metaphor: *what is in between the dots constitutes the line*. The study's central finding is that intuition is an intrinsic part of the creative process in drug discovery and thus an important organizational resource.

Benefits of a multiparadigmatic approach to creativity research

The basic argument for using a multiparadigmatic approach is to enrich the study of creativity in organizations by exploring and bridging different paradigms and thus generating new theories that may be more suitable for the study of organizational creativity in a given organization. Payne (1996: 22) argues, a multiparadigmatic approach for investigating human perception and experiences, such as organizational creativity, 'holds the promise of exploring broader, deeper and more diverse social accounts and cognitive constructions of research participants than can traditional positivist and grounded theory approaches to qualitative inquiry'. Another benefit of the multiparadigmatic approach is that different perspectives, such as collaborative management research and action research methodologies, can be more easily integrated in the research. Finally, a multiparadigmatic approach offers the potential for expanding and enriching the quality of understanding organizational creativity at different levels (e.g., individual and team). An overview of the four examples in the four paradigms is presented in Table 4.2.

The organizational setting for these examples is AstraZeneca R&D, and two smaller pharmaceutical companies (Sundgren, 2004). An important assumption is that these companies represent a spectrum of different disciplines and practices. One benefit of using a multiparadigmatic approach is that it has the potential to yield knowledge and arguments that may be relevant to the work of a variety of practitioners and disciplines within the organization. The focal point of the research was in AstraZeneca's R&D organization (including its discovery and development sectors). The overall research aim was to increase understanding of what constitutes organizational creativity in new drug development and to suggest ways in which it can be managed. This case represent four different research studies, all with different methodologies, which crossed, and used, several theoretical boundaries for the purpose of the specific research. The research was conducted as part of the Fenix Research Program. The research setting in the Fenix Program builds on two important principles: (1) practical and results can be achieved by conducting research in collaboration with industry, and (2) the researcher role must be a dual one, combining a professional role in the company with a role as a researcher in academia. As shown in Table 4.2, collaborative management research and action research is also a part of the studies (to varying degrees). *Collaborative management research* concerns two major issues. The first focuses on areas of interest for the

Table 4.2 Examples of a multiparadigmatic approach to the study organizational creativity in the case of pharmaceutical R&D

Examples (Studies)	Paradigm	Description	Research methodologies
Study I (Sundgren *et al.*, 2005a)	The functionalist paradigm	Exploring different factors for creative climate in pharmaceutical R&D.	Survey, action research, path analysis (Lisrel), based on new ways of combining different research streams of organizational creativity.
Study II (Sundgren and Styhre, 2003)	The interpretive paradigm	Retrospective study of creative projects in former Astra and ICI.	Qualitative study, collaborative research, based on the DIFI model
Study III (Sundgren *et al.*, 2005b)	The radical structuralist paradigm	Examined how different forms of performance evaluation affect the creative climate in AstraZeneca R&D. Suggest a radical change in evaluating employees.	Quantitative study, questionaire, multivariate analysis, based on situational factors (e.g. Deci and Ryan, 1980) can affect behavior related to creativity.
Study IV (Sundgren and Styhre, 2004)	The radical humanist paradigm	Exploring the role of intuition in pharmaceutical research in AstraZeneca R&D.	Qualitative study, based on Bergson's framework on intuition.

organization. The second on balance and interdependence among actors, between academic research and actual applications, between knowledge creation and problem-solving, and between inquiry from the inside and outside; the balance aims to generate actionable scientific knowledge (Shani *et al.*, 2003). Collaborative research can vary, but emphasis is placed on action research – particularly action science. In broad terms, action research methodology addresses important issues (Argyris and Schön, 1993; Coghlan and Brannick, 2001). First, the methodology involves some kind of change experimentation on real problems in organizations. Secondly, it aims to provide assistance using several iterative cycles, such as problem identification, planning, acting, and evaluating. Thirdly, action research aims to challenge the status quo from a participative perspective and is thus concerned about the intended change. *Action research* intends to contribute simultaneously to knowledge, which includes knowledge useful in academia and the creation of actionable knowledge for the client organization and social action in everyday life. Action research implies that the high standards set for developing theory and for empirically testing propositions are not to be sacrificed. Action research can then be viewed as an emerging inquiry process that is embedded in partnerships between the researcher and members of the organization where the action researcher and the practitioner will ideally jointly problematize day-to-day routines in organizations and conceive of experiments that may offer additional solutions to problems, or in other ways enhance the understanding of the problem. Effective action research therefore emanates from an experimental mindset (Styhre and Sundgren, 2005; Shani *et al.*, 2003). Furthermore, theoretical practices allow for the contextual analysis of 'thick' organizational practices and for an analysis of the multiple language games being used in organizations. Thus, theoretical practices should be examined as an important activity in insider/outsider action research (Styhre, Kohn and Sundgren, 2002).

To a large extent, methodological requirements for studying organizational creativity depend upon theoretical models (Schoenfeldt and Jansen, 1997) with a special emphasis on appropriateness. A significant part of science is to define things and then to find evidence to support the definitions. Bias comes into play when the researcher has already defined what is to be found. The point here is that using methods as a tool of knowledge construction is a double-edged sword. The sharpness is useful for digging a ditch to create a firm foundation for warranted assertions, but at the same time, one must not get stuck in the ditch – forgetting that science can afford evidence for only one of many ways of knowing and making sense of phenomena in the world. Any scientific endeavour must first define and represent (i.e., theorize) what it is looking for. Step two is to find ways to intervene in an empirical domain to investigate whether the theory captures underlying empirical realities (in the broadest sense of the term) – we represent and we intervene – in a cyclical process wherein theories are corroborated through systematic research (see Hacking, 1983).

Rickards and De Cock's (1999) paradigmatic mapping helps to explain confusions in the literature on creativity literature. There is confusion around the definition of creativity. For example, the functionalist view may 'define' creativity as the process, whereas for the interpretive perspective, it is more a matter of personal reframing. Terms such as *truth* and *validity* will also have different meanings. Thus, a multiparadigmatic approach may be a way of assessing the relative merits of each paradigm. As Rickards and De Cock (1999: 249) point out: '[T]he multiparadigmatic approach will offer considerable rewards for understanding the nature and stimulation of creativity at individual, team, and organizational levels.' The studies in this discussion (see Table 4.2) reflect a multiparadigmatic approach to the study of organizational creativity. Study I may be seen as fitting into the functionalist paradigm, whereas study II can represent the radical humanist paradigm. Study III fits into the radical structuralist paradigm and study IV is within the interpretive paradigm. In addition, these studies represent a variety of approaches in relation to the subject of research: data collection, analysis, and interpretation. Even though studies II, and IV are based on qualitative data, they differ significantly from one another. Study II is retrospective, or historical; it explores underlying aspects for organizational creativity; present interviews focus on *past* experience rather than on present activities. Study IV, in contrast, is a prospective study of present activities. Studies I and III are based on quantitative data, but study III is a retrospective study that uses an existing subset of data from a large employee survey, whereas study I is a prospective study that uses original material. However, this example argues that a multiparadigmatic approach has the potential to create tension between perspectives and offers new opportunities for understanding organizational creativity.

One concern, among several, to explain why qualitative methodology is less frequent in creativity research is validity and reliability issues. In contrast to the quantitative paradigm in which validity and reliability are more or less well defined, many qualitative researchers have struggled to identify 'how we do what we do' concerning descriptive validity and unique qualities of case-study work (Janesick, 2000). According to Wolcott (1995), validity in the quantitative arena is a set of technical micro-definitions, whereas validity in qualitative research deals with description and explanation and how well this description fits the explanation (Lincoln and Guba, 1985). In short, one way to assess validity in qualitative research is to see how credible the explanation is, and how well it fits the theory. In addition, qualitative researchers make no claim that there is only one way to interpret an event or phenomenon. For the qualitative studies in this example (studies I and II), all respondents received feedback from the analysis and made comments in their own ways. Another angle from which to view validity in studies I and II is to examine how well the organization can recognize, accept, and understand the phenomena in context. For example, the researcher's role as an insider action researcher enables a dialogue from the inside (e.g., in terms of understanding the language, images, and context), in contrast to a traditional academic researcher. Viewing validity from this perspective may even be more rigorous compared to the theoretical connection (i.e., how well the explanation fits the theory).

In summary, many methodological concerns and dilemmas can be solved using a multiparadigmatic approach to the study of organizational creativity. Much of the struggles around all-encompassing definitions and conservative thinking to create generalizability around creativity are in parallel of the so-called 'Disunity of science thesis' (Fox Keller, 2002; Dupré, 1993). This thesis deals with the central concern of the de facto multiplicity of explanatory styles in scientific practice, reflecting the manifest diversity of epistemological goals in which researchers bring to their task. Fox Keller (2002: 300) writes: 'I also want to argue that the investigation of processes as inherently complex as biological development may in fact require such diversity. Explanatory pluralism, I suggest, is now not simply a reflection of differences in epistemological cultures but a positive virtue in itself, representing our best chances of coming to terms with the world around us.' Similarly, Feyerabend's (1999: 159) advocates a pluralistic view in scientific undertakings:

> There is no 'scientific worldview' just as there is no uniform enterprise 'science' – except in the minds of metaphysicians, school-masters, and scientists blinded by the achievements of their own particular niche. Still, there are many things we can learn from the sciences. But we can also learn from the humanities, from religion, and from the remnants of ancient traditions that survived the onslaught of Western civilization. No area is unified and perfect, few areas are repulsive and completely without merit.

Expressed differently, there is a need for recognizing a methodological pluralism in creativity research.

Conclusions

Much contemporary thought on creativity is moving slowly away from psychometric perspectives towards more post-modern approaches (Feist and Runco, 1993; Runco, Nemiro, and Walberg, 1998). However, a lot of current research on creativity remains based on methodologies that are either psychometric in nature or were developed in response to perceived weaknesses in measuring creativity. In addition, creativity studies (e.g., Taggar, 2002) have generally focused on only one level of analysis at a time. This inertia of change towards a multiplicity of approaches to investigate organizational creativity is a continuing concern for researchers conceiving of creativity as what is complex, dynamic, fluid, fluxing, and therefore not easily captured by one-dimensional methodological approaches. However, methodological changes are not always easily adopted in scientific communities. Kary B. Mullis, inventor of the Polymerase Chain Reaction (PCR), a backbone analysis method in DNA analysis on biotechnology research and winner of the 1993 Nobel Prize in Chemistry, cites the conservative attitude prevailing in scientific communities as one of the major impediments towards achieving qualitative leaps in scientific work: 'Usually there are a number of powerful elders in important places that have to retire or die before things get rolling' (Kary B. Mullis, cited in Rabinow, 1996: 165).

In contrast to the mainstream methodological approach on creativity, this chapter has proposed a multilevel and combined set of methods (i.e., quantitative multilevel and qualitative) that are framed by an action research setting. Arguments for this are twofold: (i) it is important to broaden the application of psychometric methods to use multivariate methods such as path analysis; and (ii) using a qualitative approach enhances one's ability to make sense of what is observed, which can then be more easily translated and communicated into practice. So this chapter argues from a methodological standpoint that research on organizational creativity must constantly think 'outside the box' of conventional methods of research and emphasize the context in which creativity operates. In this case, this should not be limited to mixing methods or theories; it also acknowledges the importance of including, for example, action research and collaborative management research perspectives in organizational creativity research.

The following quote from Robert Fildes, CEO of Cetus, the first start-up biotechnology company to win a Nobel Prize for the PCR developed by Kary B. Mullis and his colleagues, is representative of the current situation around research on organizational creativity:

> A scientist, God bless his socks, always wants to develop a Cadillac. In the real world of products, whether it's medicine or anything else, you can bring products to the market that help a situation without necessarily being the ultimate Cadillac. That's true of drugs, of cars, of anything. I'd say, 'Come on, guys, let's get a few Fords on the way to the Cadillac, We've got to pay for the Cadillac.' (Robert Fildes, CEO of Cetus, cited in Rabinow, 1996: 155).

Part II
Practices

5
Technology and Creativity

Introduction

This chapter will critically examine the use of technology as a means for improving creativity. It will draw on the results of a study of new drug development work in the pharmaceutical industry. The study suggests that management, being the totality of practices, techniques, standard operating procedures, audits, control mechanisms, methods, and so forth, that is implemented and used in order to safeguard an organizational outcome, is criticized by pharmaceutical researchers in terms of its perceived negative impacts on research efficiency. In the case of discovery in pharmaceutical research, the outcome is a new chemical entity (NCE), a new chemical compound that serves as the basis for a new candidate drug (CD). In the case of development there are two aspects. First, the research outcome is a drug product (i.e., appropriate formulations or delivery device, and production technology for the drug). Secondly, the product containing the candidate drug is tested in clinical research activities and if it is proven to be successful in terms of benefits for the patients and is found to be without severe undesirable side-effects, the product is approved by the authorities and launched onto the market. Discovery and development pharmaceutical research is based on advanced state-of-the-art technoscience in the intersecting field of, for example, microbiology, medicinal chemistry, pharmacology, experimental medicine and drug delivery science. Since the start of the 1990s, new scientific models and methods such as high-throughput screening (HTS), computed aided design (CADD) and combinational chemistry (CC) have been used in discovery pharmaceutical research in order to raise efficiency in the early stages of new drug development (Horrobin, 2001). These new scientific screening methods represent an attempt to make use of various forms of what Bachelard (1984) calls *phénoménotechniques*, 'technologies of visualization', that enable a faster identification of NCEs.

Pharmaceutical researchers engaging in new drug development constitute what Knorr Cetina (1999) calls an *epistemic culture* and what Fleck (1979) has termed a *thought collective*. When new technologies are introduced in such professional communities, they may be treated either as an aid or as an impediment or even a necessary evil. It is a normal reaction of a community of practice to respond to events and occurrences that threaten the activities, norms and values of the community. Thus, the use of screening techniques and methods in pre-clinical pharmaceutical research represents a perceived threat to traditional practices. One of the implications for the management of organizational creativity is that there may be a trade-off between managerial control through the use of technologies and the freedom in the day-to-day work in professional communities.

The notion of technology

The notion of technology is one of the most pivotal concepts in modern society. The very idea of humanity is intrinsically entangled with the notion of technology; what makes us human, some may say, is the capacity to make use of technologies, from the simplest tool to the most advanced form of computer technology (Mumford, 1934; Ellul, 1964). The concept of technology is also an important philosophical and theoretical concept being examined and debated in numerous scholarly communities. The etymology of the notion of technology is the Greek *techne*, the art of practice of, for instance, the skilled artisan or other specialists in his or her specific field. Technology is then the *logos* of the *techne*, the speech or (more appropriately) the reason of the practice. In tribal society, the line of demarcation between technology and non-technology is easier to identify, but the lifeworld of the contemporary human being is so immersed with technologies that it is becoming difficult to exclude technology from it. 'Technology is our own nature', states the French technology analyst Paul Virilio (Virilio and Lotringer, 1997: 28). In Heidegger's (1977) treatment of technology, this inability to step outside of technology is one of its key characteristics. Heidegger is talking about the *Ge-stell*, the 'enframing' of technology, its capacity to penetrate human beings' lifeworlds. 'The essence of modern technology lies in Enframing', Heidegger (1977: 25) says. In contemporary society, the notion of technology is generally used to denote rather complicated technological systems such as consumer commodities such as television sets, computers or automobiles, but generic technologies – 'mundane technologies' in Michaels's (2000) formulation – are rather tools and intellectual technologies such as writing practices – that is, technologies that are not always recognized as such. For McLuhan (1962), the emergence of technologies of writing and printing represents a decisive moment in human history at which the human mind became better trained in thinking in terms of codification and representations: 'At any rate, with the Gutenberg technology, we move into the age of the machine. The principle of segmentation of actions and functions and roles became systematically applicable wherever desired... The Gutenberg technology extended this principle to writing and language and the codification and transmission of every kind of learning' (McLuhan, 1962: 155). What McLuhan says is that technology is not fully separated from the social lives of human beings. Instead, technology works because it is useful and makes sense to particular groups of human beings. On the other hand, technology is influencing the way in which humans perceive social reality and communicate with one another (Latour, 1991). In other words, the social is becoming technologically embedded and the technology is becoming social or situational. The science and technology studies tradition in sociology has emphasized this entanglement of the technological and the social. For instance, Wiebe Bijker (1995) provides a compelling analysis of a number of technologies and suggests that '[m]achines "work" because they have been accepted by relevant social groups' (Bijker, 1995: 270). In addition, Bijker (1995) examines technology as a form of assemblage constituted by a number of resources that have developed over time. Rather than being a unified singularity, most technologies are outcomes from distinct developmental trajectories affected by the practical use of the technology: '[A]n artifact does not suddenly appear as the result of a singular act of heroic invention; instead it is gradually constructed in the social interactions between and within relevant social groups' (Bijker, 1995: 270). In a similar manner, Orlikowski (1992, 2000) examines technology as what is shaped by practices and human engagement: 'Technologies are... never fully stabilized or "complete", even though we may choose to treat them as fixed, black boxes for a period of time. By temporarily bracketing the dynamic nature of technology, we assign a 'stabilized for now' status... to our technological artefacts' (Orlikowski, 2000: 411; see also Barley, 1986, 1990). Orlikowski (2000) thus speaks of *technology-in-practice* as what is the outcome from the integration of technological artefacts and organizational practices and as a construct that escapes an essentialist view of technology (see also Garud, Jain and Kumaraswamy, 2002; Garud and Rappa, 1994). There is thus a long-standing tradition in western thinking to examine technology and the social as being cut from the same cloth; technology is social and the social is technologically mediated. This epistemological position has developed into a great variety of theoretical positions and methodological programmes. One approach is advocated by Grint and Woolgar (1997) who claim that technology needs to be examined as a text that can be open to various interpretations and modifications. The point of departure for Grint and Woolgar is that technology *per se* is mute; it does not carry any innate qualities or objectives, but must always be examined and explored from various perspectives: 'A technology's capacity and capability is never transparently obvious and necessarily requires some form of interpretation; technology does not speak for itself but has to be spoken for' (Grint and Woolgar, 1997: 32). Since we always already speak of and about technology from some vantage point, the metaphor of *technology as text* is useful, Grint and Woolgar (1997) argue, because it allows us to escape technological essentialism and instead recognizes the contingent nature of technology. Since texts are always produced and used interchangeably – i.e., through its reading and interpretation – the text metaphor captures the notion of technology. What Grint and Woolgar (1997) are saying is that the examination of technology in practical settings needs to recognize that technology is never conclusive or determined but is always open to modifications and social influence. John Law (2002) offers an intriguing analysis of a British airplane defence system from a perspective on technology that is similar to that propounded by Grint and Woolgar (1997). Law argues that technology tends to be regarded as what objects that are once and for all determined by their materiality, their physical features. Against this view, Law argues that one needs to treat objects as 'fractional coherent objects'. That is, he feels that objects are neither singularities, single pieces of materials, nor multiplicities, assemblages of various components, rather they are objects which are altering between being unified and fragmented. Law writes:

> Knowing subjects, or so we learned since the 1960s, are not coherent wholes. Instead they are multiple, assemblages. This has been said about subjects of action, of emotion, and of desire in many ways, and is often, to be sure, a poststructuralist claim. But I agree in this book that *the same holds for objects too*. An aircraft, easy, is an object.

But it also reveals multiplicity – for instance in wing shape, speed, military roles, and political attributes. I am saying then, that the object such as an aircraft – an 'individual' and 'specific' aircraft – comes in different versions. It has no single centre. And yet these various versions also interfere with one another and shuffle themselves to make single aircraft. (Law, 2002: 2–3)

As a consequence, technological systems should not be centred, but needs to be examined as what 'balances between plurality and singularity'. As a consequence, a technological artifact 'is *more than one, but less than many*' (Law, 2002: 3). Law's argumentation is clearly counter-intuitive; we have learned that technologies are tools in the hands of and under the control of humans and that technologies are not ambiguous or elusive. Against this view, Law holds that technologies – at least technologies of the more complex kind such as the Aircraft system subject to analysis in Law's study – need to be explored as what is de-centred and possible to explore from various angles. Both Grint and Woolgar (1997) and Law (2002) are thus advocating an alternative view of technology, emphasizing the de-centred and contingent nature of technology. Technology is neither simply a ready-to-use tool, nor a deterministic system, but is rather to be regarded as what is being affected by the interaction with humans and its use in a social setting. Technology is therefore not based on its 'facticity', its brute immediacy, but on its social application and ability to be affected by other technologies and humans. Speaking with Spinoza, technology has a certain *conatus*, an ability to be affected and ability to maintain its form in and through interactions with various social entities (Spinoza, 1994: 75; Deleuze, 1988a: 99). As a consequence, technologies are never as self-contained as the dystopic visions suggests, but is always in the hands of humans who alters it in the course of action, in the very use of the technology. Therefore, technology may play a role in creative work, that is, in work that aims at providing new ideas and insights. Although technology is not inherently creative, it may be useful in the search for the creative.

Technology and representation

One specific form of technology is the ability to write, to codify a given material into a set of categories (Bowker and Star, 1999) that are shared in a community. Bolter (1991: 33) argues: 'Writing is a technology for collective memory, for preserving and passing on human experience... Eventually writing also becomes the preserver and extender of other

technologies as an advanced culture develops a technical literature.'
Writing, and its more specific form of codification, is one of the key
intellectual processes in creative work. A rich variety of technologies
and methods in use in organizations on a daily basis, such as the e-mail
system (Brown and Lightfoot, 2002), information technology (Bourdreu,
Loch, Robey and Straud, 1998) and accounting practices (Edenius and
Hasselbladh, 2002; Kreiner and Mourritsen, 2003), are inextricably
entangled with the representation and codification of conditions and
events. Gibson and Vermeulen (2003) argue that effective organization
learning – one of the key prerequisites for creative work – is the ability
to integrate three different activities, that of *experimentation*, the 'generating of new ideas', *reflective communication*, wherein 'different mental
schemes' are combined and compared, and *knowledge codification*, where
knowledge is translated into 'concrete, generalized concepts, decisions,
or action items' (Gibson and Vermeulen, 2003: 205–6). For Gibson and
Vermeulen (2003), these three processes are mutually dependent in
terms of organization learning:

> Experimentation, reflective communication, and knowledge codifi-
> cation are different actions that complement each other and,
> together, constitute learning behavior. Insights from team information
> processing and collective cognition literature (e.g., Hinsz, Tindale,
> and Vollrath, 1997; Gibson, 2001) suggests that these processes may
> be iterative rather than sequential but that they each are necessary
> for team learning behavior to occur. Hence, the three elements of
> team learning behavior are non-substitutable, that is, one cannot
> compensate for the other. (Gibson and Vermeulen, 2003: 206)

In creative new drug development work, team members need to be able
to orchestrate these three activities: 'A team will exhibit optimal
learning only if all the three elements of the learning cycle – experimentation, reflective communication, and codification – are present'
(Gibson and Vermeulen, 2003: 206). In terms of technology and creativity,
technology may be used in the experimental practices in the laboratory,
as a means for joint reflection and when codifying and storing information
for future use. What is of interest in Gibson and Vermeulen's (2003)
account on organization learning is that they point to the importance
of codifying knowledge; many accounts of creative work do not stress
the importance of codifying historical materials. However, some
researchers warn that an overreliance on codified materials may inhibit
new thinking and create an orientation towards historical facts and

records rather than one which pays due attention to future contributions. Michael Power (2004), for instance, argues that it is complicated to turn an *ex post* measure into an *ex ante* objective because there is little evidence that such management by objectives based on historical achievements will lead to higher performance. Power (2004) instead warns that codified materials – accounting practices – have negative effects on performance:

> We might say that performance measurement systems are technologies of representation which are, by virtue of their necessary reductionism, inherently defective from birth and carry the seeds of their own demise. They provide transitory managerial rationalities, myths of control, for an essentially unmanageable world. (Power, 2004: 778)

Power (2004: 779) provides an alternative to traditional accounting: 'It has been said that, at the high point of the Japanese economy, organizations did not measure cost in an elaborate way; there was rather an organizational narrative of cost in which such measures were embedded. Economic success was attributed to managing cost by "talking cost" rather than measuring it.' Adhering to a similar line of thought, Feldman (2004) argues that NASA's corporate culture is emphasizing what Feldman calls 'aperspective objectivity', defined in the following terms: 'The denial of the social context of knowledge is referred to as 'objective' knowledge or objectivity' (Feldman, 2004: 693). Feldman (2004: 692) argues that this insistence on objectivity is widespread in science-based organizations: '[T]he debate over objectivity of knowledge [is] of keen interest to the study of organizations because the belief in objectivity is central to many organizational cultures, particularly organizations that use scientific methods to accomplish their goals.' In the case of NASA, this ideology had far-reaching implications in terms of being indirectly related to the Challenger disaster examined by Feldman (2004). Feldman concludes:

> NASA actually had a great resistance to thinking in historical terms... This resistance to thinking historically was related to the belief in aperspective objectivity. Historical knowledge is based on the uniqueness of the data. Generalizing across these data is limited because of the unique detail of the record. Historical understanding is understanding of human experience and its changes over time. A culture dominated by the belief in aperspectival objectivity, on the other hand, focused on finding general explanations for synchronic

relationships in empirical performance. This is why the engineers saw nothing wrong in looking at only the most recent flight: they imagined they could attain complete knowledge by grasping the causal relationship between parts. What we can learn from this is that the rationalized empiricism of the culture of aperspectival objectivity developed a set of abstractions for understanding that left out or ignored parts of reality relevant to its own goals. This added to their tendency to underestimate flight risk by limiting the amount of data they considered enabled them to exaggerate flight safety. (Feldman, 2004: 713)

There is a clear moral in Feldman's analysis: no single source of knowledge may be used to discredit others' forms of know-how. Knowledge codification is of great help in creative work, but should not be used to exclude alternative forms of thinking. In creative work, technology may be used for a variety of purposes. One of the most important functions is to codify and store information provided from laboratory research activities. For Gibson and Vermeulen (2003) such codification is an important part in a learning culture while some researchers offer a more complex view of codification, suggesting that creative work may in fact be inhibited by a single-minded emphasis on a certain performance measure. In either case, technology plays a key role in providing the means for such practices of codification. Thus, technology is again neither a curse nor a blessing *per se*, because it can be used in a wide variety of ways. The most important point to be drawn from the analysis of the use of various forms of technology in creative work and in new drug development work is that technology is not detached from politics and practices but is an integral component in organizational practices. Technology is then neither 'good' nor 'bad' per se, but always need to be examined within its social setting and practical functionings.

Technology and creativity

The emergence of modern advanced science and biomedical science is intimately related to the development of increasingly advanced technologies for the collection, analysis, storing and sharing of data and information (e.g. Howard, 2000). Without, for instance modern computer technology, a great deal of modern science would be simply inconceivable: 'No science without technology, without machines', as the French philosopher Michel Serres (1995: 15) puts it. The lifeworld inhabited by the contemporary scientist is filled with a broad variety of advanced technologies and equipment. As a consequence, the knowledge employed by the laboratory scientist in the pharmaceutical research is best regarded as what Lanzara and Patriotta (2001) refer to as an *assemblage*, a set of heterogeneous resources comprising technologies and theoretical frameworks. Lanzara and Patriotta (2001) write:

> Rather than a discrete commodity, organizational knowledge could be better pictured as an 'assemblage' subject to continuous transformations and reconfigurations. It is an assemblage precisely because it is the outcome of controversy and bricolage, resilient as a whole but subject to local disputed, experiments, and resembling... An assemblage is neither a unity not a totality, but a multiplicity, a collection of heterogeneous materials that are mutually but loosely interrelated. In other words, the notion stresses the importance of relations over the elementary parts, i.e., what goes on 'between' the part (Cooper, 1998, p. 112). In this regard, what makes knowledge distinctive is not the discrete collection of commodities, but the nature of the assemblage and, we should add, the making of the assemblage in time. An assemblage is an evolving artifact and it is unique because it springs out of unique history. In summary, the notion of assemblage emphasizes the pasted-up, path-dependent nature of knowledge systems and reinforces the definition of knowledge as a phenomenon in the making, which eventually make sense in the retrospect. (Lanzara and Patriotta, 2001: 964)

Although Lanzara and Patriotta (2001) speak of knowledge as an assemblage in terms of being a mixture of know-how and technologies, it is important to emphasize that all technoscience is in addition to its technological constitution is also highly political and social in nature. In the pharmaceutical industry, new disciplines such as bioinformatics, proteomics and genomics have emerged.[29] These disciplines aim to provide new opportunities to simultaneously investigate the

---

[29] Genomics is the large-scale use of small molecules to study the function of gene products. Proteomics, a branch of functional genomics, is the large-scale analysis of polypeptides during cell life; its purposes are to catalogue proteins that our genes encode and to decipher how these proteins function to direct the behaviour of a cell or an organ. A technology like bioinformatics is the cross-discipline of computer science and biology; it seeks to make sense of information from the human genome, to find better drug targets earlier in drug development (Hopkin, 2001; Howard, 2000).

structures and function of very large numbers of genes; these opportunities have generated a lot interest and activity across the biotechnology and pharmaceutical industries (Ezzell, 2002). In his intriguing analysis of genomics research in France, the anthropologist Paul Rabinow conceives of genomics as a shared field of interests in which heterogeneous resources are brought together and mutually reinforce one another. Rabinow writes: 'French DNA is about a heterogeneous zone where genomics, bioethics, patients groups, venture capital, nations and the state meet. Such a common place, a practical site, eruptive and changing yet strangely slack, is filled with talk of good and evil, illness and health, spirit and flesh. It is full of diverse machines and bodies, parts and wholes, exchanges and relays' (Rabinow, 1999: 4). Here, creativity in the field of genomics is not only a matter of being able to bring together know-how and relevant technologies, but also an effect of the entrepreneurial capabilities of the researcher, i.e., the ability to attract venture capital, political support, and legitimacy within the field (see also Rabinow, 1996). In this view, the creative researcher is not only skilled in laboratory work and knowledgeable in particular fields of research, but is equally in possession of the *savoir-faire* of the entrepreneur.

Even though creativity is not synonymous with knowledge, knowledge is a *sine qua non* for creativity, a qualifying factor for being able to contribute creatively to a particular field. One particular function of modern laboratory technology is to enhance perception, that is, to function as what the French technology analyst Paul Virilio (1994) calls a 'vision machine', a machinery that enables faster and more adequate 'logistics of perception' (see, e.g., Traweek, 1988, on the use of laboratory technology among physicists). Even though technology plays a decisive role in laboratory work, technology is by no means an enclosed and ready-made piece of machinery; rather it is instead a highly malleable resource in the hands of the laboratory researchers (Lynch, 2002). Jacques Ellul (1964: 6) formulates the relationship eloquently: '[W]hen a technique enters into every area of life, including the human, it ceases to be external to man and becomes his very susbstance. It [technique] is no longer face to face with man but is integrated with him, and it progressively absorbs him.' Technology is thus affecting the organization and its work routines. On the other hand, technology is also adapting to the environment in which it is located. Studies of the use of technology in organization suggests that there is a mutual adaptation between technology and organization (Hayes and Walsham, 2003; Edmonson, Bohmer and Pisano, 2001;

Woicehyn, 2000; Noon, Jenkins and Martinez, 2000; Goodman and Sproull, 1990: Leonard-Barton, 1988).

In pharmaceutical research, with its central aim of providing new drugs which offer medical benefits, the notion of perception is of central importance. The notion of vision is perhaps the single most important metaphor for scientific discovery in the field in terms of detecting new chemical compounds that have promising properties given the focused indication. In laboratory work, synthesis chemists are screening a large number of molecule structures that may be the basis for a future drug. As a consequence, one of the areas in which technology may enhance the performance of the work is to offer 'vision machines' that accelerate the discovery processes. In this chapter, the pharmaceutical researchers' view of such vision machines or technologies of detection is critically examined. As will be pointed out, pharmaceutical researchers are not ready to offhand accept the entrance of new laboratory technology because it effectively alters the laboratory practices and the perceived status of the laboratory worker. For instance, Thomke and Kuemmerle (2002: 631) write: 'Field interviews reveals that traditional chemists felt threatened by the new technology [High throughput screening] that appeared to automate many of the tasks that they had so carefully learned and refined over many years.' Barley, studying the use of CT scanning technologies in two hospitals, emphasizes the social embeddedness of technology:

> Technologies are depicted as implanting or removing skills much as a surgeon would insert a pacemaker or remove a gall bladder. Rarely, however, is the process so tidy. Events subsequent to the introduction of a technology may show that reputedly obsolete skills retain their importance, that new skills surface to replace those that were made redundant, or that matters of skill remain unresolved. In any case, groups will surely jockey for the right to define their roles to their own advantage. (Barley, 1990: 67).

This concluding remark implies that the arguments of students of technology such as Bijker (1995), that technology is affecting social relations but is also social per se, is worth taking into account in empirical research on the use of technology in creative work. Even though the vision machines of high-throughput screening are rational tools in the quest for new chemical compounds, the social reality of laboratory scientists is informing and affecting the reception of the new technology. In other words, there is no hermetically sealed technology that escapes influence from the social setting in which it is located; rather, all technologies become useful through being aligned with dominant beliefs, ideologies and practices. In other words, technology can contribute greatly to creative work but it needs to enter the community of scientists in a gentle manner, heedfully interacting (Weick and Roberts, 1996) with pre-existing practices. Kallinikos (1996: 53) expresses this idea eloquently:

> Rather than being simply a variable or isolated set of factors that impinges upon some aspects of organizational life, technology can be said to represent an integrated system of perception and interaction that define basic forms of everyday organizational activitity. Technology is more than simply a means to the predefined ends of innovation and effective production.

Technology is bound up with social practices; similarly, social practices are constitutive of technology. The technological is socially embedded.

High-throughput screening and new drug development

Pharmaceutical research including discovery and development is based on advanced technoscientific laboratory work. The search for NCEs is taking place in distributed knowledge systems where a number of different expertise are cooperating. NCEs are the outcome from the joint efforts of, for example, medicinal chemists, biologists, physicians and pharmacologists. Laboratory work and scientific work are dependent on the commitment and knowledge of those who are participating. Almost all of the interviewees claimed that they enjoyed working with new product development and that they thought the work was both exciting and rewarding. The researchers were focused on providing opportunities for new better pharmaceutical products that could provide a better therapy for the patients and beneficial for the society. The interviewees also emphasized that the search and identification of NCEs were dependent on the ability of the participating researchers to cooperate and share knowledge and skills. Thus, the researchers established a community of practice that shared a vocabulary, a deep understanding for the nature of the challenge to develop a NCE and a set of practices, techniques, norms and values that were employed in the day-to-day activities and operations. All communities of practices operating within a specific research programme therefore become *homogeneous*; there are mechanisms establishing various shared view and objectives that are not subject to discussion or criticism. Homogeneous social formations are always based on a number of taken-for-granted assumptions and ideas. However, the top management of AstraZeneca represented a heterogeneous component of the new product development activities. The homogeneous scientific community undertaking laboratory research did not have the same objectives or long-term strategies as top management.

Therefore, top management's decisions were seen as an impediment or threat to the scientific activities. Pre-clinical researchers were not too affirmative towards the idea of managerial interventions in their day-to-day work. Management was seen as something fundamentally different from leadership, which the researchers were more willing to embrace. One of the interviewees argued:

> Well, I think there are some very, very creative people, some very able people but it's the managers that are the problem. I wrote a report . . . you could summarize it by saying there some really brilliant people here doing some super science. Pity about the management! And the management had been doing the wrong things and was inconsistent in its approach. I believed that . . . if you have them committed you have them – and . . . then people get really quite excited and will move mountains to achieve what you want them to achieve. If they don't believe in the value of what you're doing and they don't actually think you understand what you are trying to achieve then they're not going to put themselves out very much, why should they? Because they're not actually committed, they don't believe in it themselves. So I think it's the leadership that needs to be provided in science and a consistency of leadership and there's also a communication of why we're doing it. Making people feel that what they are doing is important. And there's a tendency in some parts of the organization to use data and information as power. I'm not going to let everybody in my group know what's going on. In fact, I don't know what's going on in their minds but they are not going to know, 'cause that's my job. I want the power to know what's going on. (AstraZeneca, Manager, Discovery)

In brief, the pharmaceutical researchers wanted to run their research projects without any detailed monitoring or intervention from the management. On the other hand, management wanted to ensure that the researchers provided the company with the right (i.e., drugs with a promising market potential) candidate drugs that were complementary to the company's product portfolio and its long-term strategies. The researchers wanted the greatest intellectual and scientific freedom possible while top management wanted to maintain a reasonable control over research activities. Therefore, some of the researchers felt themselves to be perceived as a potential threat to top management in terms of being the source of competitive advantage in the company. Creativity, understood as the ability to bring forward new NCEs and develop competitive drug delivery systems, was at times conceived of as something ambiguous: on the one hand, it was seen as the source of competitive advantage, on the other hand as something threatening. One of the researchers claimed:

> Well, creativity means several things to me, I mean, the first thing, creativity means is the ability to tell a story, and by that I mean the ability to pull together all the information and put it into a form that other people can understand, the second thing that creativity means is to me, is not accepting the status quo, so if someone did a trial in a certain way, because I am a physician, in a certain way yesterday, creativity is finding a better way of doing that trial, not just repeating what people did yesterday, and, and creativity for me as well, is challenging other people, you know you have to say, are you sure about that, have you really thought about that. I think we would probably have a common agreement on the definition of innovation, you know the small incremental steps, to gradually improve things, I think most people, not just in this company but in any company, can get quite scared of creativity, it's dangerous, it's threatening, it's nothing wrong with that, but it you know it says sorry we want you to think a completely different way, not just to make this way slightly better, and that's quite difficult to know organization in R&D and its 10,000 people and it's very difficult to suddenly switch everything round or, or, or whatever possible. (AstraZeneca, Manager, Development)

Creativity is thus both a key to success and a capacity that can be used to pursue political means in the organization. To the researchers creativity is a resource for the scientific work while management aimed to control it. One such means of control was the reward system that was claimed to emphasize individual initiatives rather than team-based cooperation:

> The hierarchy is destructive for information. Information is the backbone of a creative environment. If that prerequisite, the access to information, is absent, then there will be very little creativity... Unless you have a good way of distributing information, then only a few persons at the top will be able to present new ideas... All this is reinforced by that damn reward system: there is nothing but a dog-eat-dog mentality where you just care for yourself. It's all very destructive. Unfortunately, top management does not recognize the problem, because they are not aware of the basic mechanisms of scientific activities. (AstraZeneca, Researcher, Discovery).

The ambiguous and somewhat tense relationship between managerial activities and research activities produced a sceptical attitude towards the use of various laboratory screening technologies and methods. One of the interviewees lamented these tendencies in the pharmaceutical industry:

> One of the problems today is that we have become very process-driven and we rely very heavily on high-throughput screening at the moment. I think that is counter-intellectual, I don't like high throughput screening. I feel we have to do it, but the sooner we get rid of it, the better. Because it's throwing stuff against the wall and hope that something sticks and saying yes, that's worth picking off the wall. We didn't do that in the old days. There was no possibility doing it, so you have to exercise your brain in what you are going to make. (AstraZeneca, Manager, Discovery)

These advanced technoscientific forms of trail-and-error screening of NCEs were conceived of as being a managerialist approach to strategic opportunities. In traditional pre-clinical laboratory research, researchers were responsible for finding NCEs though established standard operating procedures and routine work. In these research settings, there was very limited influence by managerial interests. Management rather had to make use of the candidate drugs that were provided by the laboratory research activities. In increasingly competitive markets, traditional laboratory research was perceived as being too slow and inefficient and therefore various technoscientific techniques and methods were used. In other words, managerialistic objectives and practices penetrated the homogeneous pre-clinical laboratory community of practice. This caused a number of negative responses among the researchers:

> My earnest hope and belief is that HTS [high-throughput screening] is a [vanishing] phenomenon and that HTS will be relegated to a minor activity in ten years time. By that time we will have structures of most of the major proteins and that will be modelling 'in silico'[30] and making libraries to test hypothesis. (AstraZeneca, Researcher, Discovery)

Another interviewee argued that the proclivity towards investing in technoscientific technologies represented a managerialistic ideology where a convergence towards pre-existing, 'fashionable' practices was highly favoured. Top management was claimed to be willing to invest in such technologies and routines because other companies did that:

> I have never heard anyone within this company ask the question why we were so successful in the end of the 70s and the early 80s. Why were we so creative, how do we manage to produce that many candidate drugs in such a small organization? What were the success factors and how should we reproduce them? Rather than trying to do what we were good at, we were more concerned about what others did. And the grass is always greener in your neighbour's garden: we saw companies investing in bio-technology and micro-biology and the latest fad high-throughput screening, molecular modelling, drug design and all that. I ended up in a situation where I did things because top management should think I was modern up to date and not just an old reactionary, right. Sooner or later you realize that it does not work. You cannot do things you do no believe in. (AstraZeneca, Discovery, researcher, Sweden)

Top management here represents a mode of thinking that in a sense excludes scientific activities. Management was oriented towards control

---

[30] In-silico refers to 'computational' denoting the combination of advanced mathematical and computational methods, techniques and simulation tools (bioinformatics/biocomputing) with experimental (in-vivo) and clinical knowledge, particularly during the drug discovery phase, with the aim of better understanding and treating the disease.

and adoption to external changes in the market and in the industry. Scientific activities aim to create new findings and new results. On the one hand, control, structure and organization are praised corporate virtues; on the other, intellectual freedom, creativity and novelty are favoured. Hence the scepticism toward a managerialistic ideology and agenda. One of the respondents recalls previous attempts at managing the activities:

> Take any other industry, there you're asking yourself all the time what are the reasons for success, for being competitive. I recall when I was a young researcher when PharmaCorp invested in fish protein and diverse activities and God knows what, and they rejected investments in everything related to pharmaceutical development. The years passed by and still pharmaceuticals were the source of income for the company. All those other ventures yielded pure losses, like Ericsson's mobile phones, right [in the beginning of the new Millennium]? If you're lucky, those mistaken ventures sink into the sea and then you can invest in pharmaceutical research anew. Sure, you can invest in new businesses, but one mustn't do that without being aware of your key success factors. (AstraZeneca, Researcher, Discovery)

For the pharmaceutical researchers, the very core of the industry and the company was the ability to undertake advanced state-of-the-art scientific work. Without this core competence, the company would be susceptible to all kinds of management fads and institutional isomorphisms that would seriously damage the long-term competitive advantage of the company. The main message to emerge from the researchers was that the company should stick to its key objectives and refrain from diversification.

Other forms of technology in new drug development

The laboratory researcher and other workers in the development of new drugs in the pharmaceutical industry inhabits a domain penetrated by all conceivable technologies that in various ways support the sharing of information and distribution of knowledge throughout the organization. A continuing concern for chief information officers and other individuals responsible for maintaining the information infrastructure in pharmaceutical companies is the effective sharing of adequate information. Even though such information-sharing systems are intelligently designed and enable advanced forms of data and information search, research suggests that it is difficult to exploit full potential of such systems. A number of factors contribute to the relatively modest use of information technology in new drug development work. A survey study conducted in 2004 at AstraZeneca, offered a number of responses on the questions of information sharing and the learning culture. For instance, one of the development researchers argued that the abundance of systems and the policing of such systems constituted a practical challenge in day-to-day work:

> There is clearly a wealth of information on the intranet, but discovering it is more 'accidental' than not. Use of unique login IDs and passwords for so many sites is another impediment to finding/sharing information. (AstraZeneca, Researcher, Development)

In a similar manner, another researcher from the development organization argued that information is not easily found because there are databases that are not open to everyone:

> GEL [an electronic library] needs to be open to viewers. I have found that I need information to do my job and do not have read access to information. Then getting the access takes too much time. There are a lot of secret folders that are not available to viewers and this prevents people from being able to do their job efficiently. For example, I had to wait two days just to have access to a document that I needed so I could create a higher-level document. (AstraZeneca, Manager, Development)

The inertia of information-sharing systems, dependent on various safety routines such as authorization, is often a source of frustration rather than an aid to innovation. Creative work is supposed to be reinforced by swift information sharing, but in many cases the 'logistics of information' poses a real problem in the work. As another development researcher points out, the databases are not always user-friendly or easy to find:

> I am sure systems are available. The issue is that they are not coordinated, not user friendly, or even easily identifiable. There is a real need to ensure that information management is supported by systems which support the needs of users can be shown to be to their benefit and thereby enhance sharing of and access to information. (AstraZeneca, Researcher, Development)

Another researcher made the following observations:

> Sharing information is only useful if key words are provided to enable intelligent searching. Else, too many hits are red herrings or provide information overload. Sharing information involves greater work for the author in this respect. Intranet sites are underutilised and set up poorly – in effect making them glorified filing cabinets with no filing system! We do not make full use of the web functionality already available to us. (AstraZeneca, Researcher, Development)

In addition to a number of databases storing information for future use, new drug development work is making use of a variety of standard operating procedures (SOPs) and other forms of rule-governed practices. For the individual researcher this implies that creative work is always already determined by a number of practices and routines:

> It is sometimes hard to have new ideas, as we have so many SOPs, guidelines, specified communication routes, etc. You feel like you have to stick to the rules, even if they are stupid, and make things take longer time. (AstraZeneca, Researcher, Development)

New drug development work also makes use of more mundane office technologies such as e-mails and regular meetings. These are also supposed to enable creative work, but are in many cases regarded as additional sources of stress in terms of continuously providing even more information:

> It is very difficult to stay on top of e-mails. Because of the lack of time, I very rarely go to intranet sites to look for information. I rely on my contacts with functional representatives to get the information that I need. (AstraZeneca, Researcher, Development)

Although the survey results indicate that there is a general distress over the poor *de facto* functionality of databases, e-mail systems and other information technologies, most new drug development researchers would agree that without these tools, the whole process would be much slower and more labour-intensive. Still, the belief in information technology as some wholly transparent and efficient system that only provides information when requested and that does not steal any time from other activities remains a utopian one (Valentine, 2000). For most new drug development researchers, information systems are helpful but they are also sources of frustration in terms of promising more than they can deliver. The interaction between man and machine is then still complicated in terms of failing to become a smooth interaction between the two systems. Creative work is still awaiting information technology that provides information without demanding valuable time from the researcher.

The limits of technology

Alluding to Heidegger's (1977) famous dictum that 'science does not think' (in relation to philosophy), one can similarly say that management (as being a set of practices and techniques) does not think: It does not provide local solutions to local problems but implements wide-ranging, conclusive managerial models where other solutions are expected. To follow Bourdieu's (1996) distinction between heterogeneous and homogeneous fields, one can say that management implement heterogeneous actors, i.e., managerial techniques and practices in homogeneous fields – pharmaceutical research communities – and thus erode the consensus within a community of practice through imposing complementary agendas and concerns. The technoscientific field within discovery research in particular, but also with development research, is thus exposed to different objectives and goals than it has been used to traditionally. Discovery researchers have been trained in providing NCEs and CDs to the clinical research organization, not to take care of managerialistic research technologies.

If pharmaceutical researchers reject managerialistic models aimed at competitive advantage, what do they then embrace? One straightforward answer is *leadership*. Whereas management represents the day-to-day activities and practices and continuous control of the laboratory operations, leadership is the practices that enables for a development of existing activities. Professional organizations (see e.g., Mintzberg, 1983; Reed, 1996; Mueller and Dyerson, 1999) are generally claimed to be different from non-professional organizations. Professional organizations need to provide opportunities for its employees rather than simply to impose control. In the case of AstraZeneca, the pre-clinical researchers were willing to accept or were even positive about the idea of good, committed leadership of laboratory research activities. Traditionally, AstraZeneca's great successes had all been based to some extent on a clear vision of the activities and a long-term objective. However, to pharmaceutical researchers, the difference between leadership and management within AstraZeneca was indistinct or vague. Top management's will to provide leadership often resulted in further managerial activities. Therefore, the pharmaceutical researchers were characterized by, to make use of a new concept, a certain *saudade*. *Saudade*, generally translated as 'yearning' or 'brooding', or a 'variety of a state of anxiety tempered by fatalism', is a Portuguese concept denoting a feeling of nostalgic melancholia and a loss of a successful past, a longing for a time that has passed and will always overshadow the possibilities of the future. *Saudade* is claimed to be a national characteristic of Portugal, a small and geographically peripheral (in relation to Continental Europe) nation that once was a great colonial power and a nation of explorers of the world. The pharmaceutical researchers loss of leadership and a clear understanding of the laboratory technoscientific activities among top management entailed *saudade*, a slightly nostalgic feeling of a successful past whose equivalent is never to be seen again, among the researchers. The emergence of detailed managerial technologies and practices represent a radical break with traditional pharmaceutical research within Discovery and Development. Moreover, those managerial technologies can never fully replace and substitute the leadership and community of practice-based activities of the 1970s and 1980s. Management thus denotes the loss of certain technoscientific virtues.

Technology as a form of control

Laboratory work in new drug development activities is inextricably entangled with the use of various forms of advanced technologies. However, the use of technology in creative work is not uncontested. Studies of work in other settings such as in manufacturing work show that technology may serve to monitor and supervise workers (McGail, 2002; Ball and Wilson, 2000; Sewell, 1998; Kidwell and Bennett, 1994). In other words, there are technologies and practices that are helpful in the day-to-day operation of the laboratory, while other technologies and practices are regarded as a threat to professional autonomy (see Doolin, 2002; Humphreys and Brown, 2002; Kitchener, 2000). As a consequence, it is not the technology per se that is contested; rather, it is the underlying managerial regimes of control. Kelemen (2001: 2) distinguishes between four different forms of control:

> Direct control refers to coercive mechanisms by which individuals are made to do things they would not do otherwise (e.g., supervisory control). Technical control is a form of indirect control which relies

on the use of technology (e.g. the assembly line) to get people to conform. Bureaucratic control refers to the internalization of rational rules and routines by organizational selves. Discipline is a form post-bureaucratic control which draws significantly on all these previous forms of control and yet it is, to a certain extent, different in that it appears not to control but to offer a high degree of individual autonomy at work.

Technical control is often not regarded as a form of control per se but rather as some direct consequence of the use of technology. When the use of technology is becoming similar to that of direct control or when technology is intervening with the standard operating procedures of the laboratory technicians, it may be regarded as problematic. In the case of HTS, the new technology directly intervened with the laboratory researchers' expertise and professional skills, turning the laboratory work into an automated screening process. HTS is then – as opposed to for instance intranets that are designed to support rather than displace laboratory scientists' skills – posing a threat to laboratory researchers. Newell, Swan, Scarborogh and Hislop (2000: 103) write on Intranets: 'Intranets are a de-centred technology...that is, loosely coupled systems with no core or essential characteristics or significance but rather multiple and distributed meanings and actors.' HTS, on the other hand, is not loosely coupled or de-centred but is rather located at the centre of operations; it is a central technology rather than being part of the laboratory infrastructure. The disregard of the HTS technology is then founded in the unwillingness to reduce a sophisticated laboratory expertise to the level of machine operations. Since laboratory work is a domain of creative work, laboratory workers are not willing to turn their work into some machine-based process. In Strati's (1999: 176) words: 'When we study people's creativity...we observe organizational forms very different from those to which we have been accustomed by the dominant Taylorist and Fordist models.' The implications for the management of creativity and organizational creativity is that professionals are not very eager to offhand accept technologies that in any direct, indirect or symbolic way are challenging their status as legitimate experts in a specific domain. Professionals may be willing to recognize the importance of technology when there is no alternative or when the technology is supporting the work practices but when the technology is actually introduced as an alternative to the traditional expertise it is becoming contested. Creative work is not possible to separate from emotionality (Carr, 2001; Huy, 1999; Fineman, 1993) because the professional worker is highly committed to their work; new technology is often treated with scepticism since it is emotionally disturbing to observe the entrance of new technologies into the workplace, the realm of creative work and, to some extent, a realm of self-fulfillment.

Conclusions

Management as a practice and a scientific discipline is fundamentally based on the Enlightenment ideal of understanding the world as a rational closed system. Thus, managers are prone to accept managerial technologies and practices that structure and organize a complex or complicated perceived reality. Management activities can therefore do good as well as they can do harm; they can enable things just as they can produce roadblocks and impediments. In pharmaceutical research, in essence a professional, expertise-based and knowledge-intensive organizational activity, managerial technologies and practices are not always praised and well received. They are rather seen as technologies that are based on their own innate rationales which are removed from the pharmaceutical research agenda. The will to undertake certain organizational activities more efficiently and in accordance with specific managerial principles or dominant ideas may at times be more focused on control and audits than on increased output. In future, the pharmaceutical industry needs to address the differences between management and control on the one hand and leadership and the production of creative work on the other in order to sustain and reinforce its creative abilities in terms of producing new chemical entities and pharmaceutical products.

Even though technology is a *sine qua non* for modern new drug development practice, the reception of new technology among the professional community of laboratory researchers is dependent on the perceived underlying rationale for the new technology. The management of organizational creativity therefore needs to take into the account the influence of and reception of technology in different communities. New technologies are rarely regarded as solely a blessing, but are often treated as something that consume valuable time rather than releasing time. As a consequence, one should abandon simplistic technology strategies and rosy images of technology as some kind of universal solutions to all kinds of managerial evils and pursue a technology strategy that effectively supports and enhances creative work in organizations. As several writers have emphasized, technology does not exist in a social vacuum but is rather moulded in the day-to-day practices in laboratory and workplaces. Hence the importance of a deliberate and thoughtful technology strategy in firms relying on their creative competencies.

6
Intuition and Creativity

Introduction

This chapter will address the human faculty of cognition. In management studies and organization theory, managers and co-workers' cognitive capabilities have been a source of investigation since the end of the Second World War when Herbert Simon introduced the behavioural theory of decision making. In Simon's theory, decision making in organizations is determined by what Herbert Simon (1957) calls the *bounded rationality* of managers and other decision makers. This insight has significant implications for the functioning of organizations. For instance, rather than being concerned with optimal solutions to problems, organizations are merely satisficing their decisions, reaching 'good enough' decisions given the degree of ambiguity and chance in a certain situation. Simon's research programme has been a major source of influence in management studies and organization theory after the Second World War. When speaking of creativity, the notion of intuitive thinking is what is of interest in this chapter. As a subset of cognition, intuition remains one of the least exploited cognitive faculties of human beings in the management literature. Moreover, the role of intuition receives little attention in the literature on organizational creativity. This chapter describes a study of the role of intuition and its implications for organizational creativity within pharmaceutical research. The study applies the French philosopher Henri Bergson's philosophy of intuition wherein knowledge is separated through use of ready-made concepts, and intuition is the ability to think 'in-between' these concepts – to think in-between points of existing knowledge. The study is based on a series of interviews with employees in pre-clinical research (Discovery) in AstraZeneca. This chapter concludes that intuition is a resource that facilitates new drug development. Pharmaceutical researchers perceive the roles of intuition and creativity as intertwined in ground-breaking scientific contributions. But intuition is a contested construct in the organization because it is in opposition to reductionistic and analytical forms of thinking which are highly prized in much of the new drug development literature. Nevertheless, Bergson's philosophy may form a foundation from which to explore intuition and its relevance for organizational creativity.

The notions of cognition and meaning

In many cases management studies and organization theory are implicitly addressing the cognitive abilities of practicing mangers and co-workers in organizations. Most managerial practices and undertakings in organizations are dependent on different forms of human thinking. However, only a subset of management studies explicitly draws on a theoretical framework that explores the notion of cognition. For instance, in the Academy of Management, the most important management research association in the USA, one division is dedicated to the topic of management cognition and a research community attending conferences and publishing papers on management cognition has been developed. The notion of cognition is derived from psychology and is defined by Leon Festinger (1957: 3) as follows: '[A]ny knowledge, opinion or belief about the environment, about oneself or about one's behavior.' This is, to say the least, a broad and rather elusive definition, but it nevertheless points to certain characteristics: it stresses the three different levels of thinking (knowledge, opinion and belief) and points to the relationship between the self and its environment. According to Bruner (1990), research in psychology moved from behaviourist to cognitive explanations in the 1950s. Bruner, a noted researcher representing this new orientation, talks about this change as 'the cognitive revolution'. Rather than assuming that human beings are rationalist automata responding to external stimuli such as threats or encouragement, cognitive theories postulated that humans are guided by their cognitive capacities, the ability to think and conceive of alternatives before taking action. The rationalist and instrumental theory of behaviourism thus fell from grace and was displaced by a cognitive model of humans. In sociology, a similar movement from functionalist or system models in the tradition of Talcott Parsons became criticized by constructivist sociologists and phenomenologically oriented researchers. Sociological schools such as symbolic interactionism and ethnomethodology, represented by sociologists such as Charles Cooley, Herbert Mead, Alfred Schutz and Harold Garfinkel, emphasized notions such as 'meaning' and 'understanding' in the sociological vocabulary. Management studies, a discipline closely related to disciplines such as sociology, psychology, and political science, moved in the same direction. At the end of the 1950s and during the 1960s, James March, Herbert Simon and Richard Cyert further developed the theories of decision making, placing cognition as a key parameter of organization performance. In 1969, Karl Weick published the seminal work *The Social Psychology of Organizing* in which concepts such as *sense-making* and *enactment* further advanced the idea that cognition is of great importance when understanding organizational activities. In a series of papers, Weick has elaborated on these ideas and persuasively argued that the way in which people think about their work and their lives have far-reaching implications for management practice (see e.g., Weick, 1996, 1995). Researchers such as Gioia and Chittipeddi (1991) have suggested that individuals in organizations do not only make sense out of their work life experiences but are also given sense through managerial practice and leadership practices. The notion of *mapping* has been advocated by Huff (1986) and Ambrosini and Bowman (2002) as a tool to visualize shared images of reality in organizations. Greve (1998) and Greve and Taylor (2000), and Augier and Thanning Vendelø (1999) have respectively explored the notion of cognition in studies of innovation and knowledge management in network organizations. In more specific terms, the notion of cognition is often operationalized in management studies as 'meaning'. Meaning is a notoriously fuzzy concept that appears in various definitions, but is still regarded as what is creating a sense of coherence for individuals. In order to provide the notion of meaning with a definition we turn to the German sociologist Niklas Luhmann. Luhmann writes:

> 'Meaning' is fundamental to human experience and action: It is constitutive of time and history to the extent that it enables us to experience the selectivity inherent to all aspects of social life. In other words, every event in which meaning plays an essential role takes place within a horizon of other possibilities. (Luhmann, 1982: 293)

Luhmann continues: '"Meaning" may be defined as the conjunction of a horizon of possibilities with selection of choice. "Meaning" in this sense makes especially effective forms of selectivity available to systems' (Luhmann, 1982: 345). Meaning is here the outcome from making a selection of possible choices. Elswhere, Luhman continues: 'Meaning...is actuality surrounded by possibilities. The structure of meaning is the structure of this difference between actuality and potentiality. Meaning is the link between the actual and the possible; it is not one or the other' (Luhmann, 1990: 83). What does this all mean? One may say that meaning is the effect rather than the point of departure from human actions; humans make choices from a set of perceived differences and those choices are made legitimate through various forms of sense-making. Meaning is then the sum of all those activities aimed at making sense. In a less technical language, the British sociologist Zygmunt Bauman (1999: 96) writes: 'Human praxis, viewed in its most universal and general features, consists in turning chaos into order, or substituting one order for another – order being synonymous with the intelligible and meaningful.' Such human praxis has been grist to the mill in management research. A great variety of studies have examined how organization members make sense out of their everyday work life experiences through various practices and undertakings, for instance, by telling stories (Gabriel, 2000) or gossiping (Kurland and Pelled, 2000). Organizations are domains where meaning is produced through the social interaction of individuals. Daft and Weick (1984: 293) put it rather succinctly: 'To survive, organizations must have mechanisms to interpret ambiguities and to provide meaning and direction for participants.'

However, the notion of cognition is a wide-ranging construct that comprises a variety of cognitive and perceptual capabilities that serve human action in different ways. When speaking of creativity, there are a number of alternative perspectives on the relationship between creativity and cognition. In this chapter, a specific form of cognition, namely what is referred to in the literature as *intuition*, will be explored as an important component of creative work. In new drug development processes, there is an immense reliance on scientific practices and methods derived from disciplines such as chemistry, biology, pharmacology, and a number of sub-disciplines within the life-sciences. Although these scientific procedures are the foundation for creativity in new drug development, a sole emphasis on instrumental rationalities and functionalist operations is only partially capable of explaining superior performance in new drug development. Residual categories such as chance and luck – labelled as 'serendipities' in scientific work (see, e.g., Roberts, 1989) – is one form of explanation that stretches beyond the rationalist domain. In a similar manner, the notion of intuition is helpful in capturing scientific procedures in their making, at the point in time where theoretical frameworks, laboratory machinery and equipment, and empirical results are not yet stabilized and verified, but is still in a situation wherein they are continuously being altered and moving in the direction of the equilibrium – that is, the position wherein theory and empirical results are being co-aligned and made conclusive. Intuition is then the researcher's capacity to draw conclusions on meagre or partial results, formulating hypothesis on basis of a not yet finalized empirical material. The ability to draw conclusions on scientific results *en route* and to find shortcuts is one of the most important skills when operating under time and money constraints. In pharmaceutical research, scientific work is always restrained by short-term goals and financial concerns. As a consequence, laboratory scientists develop their ability to operate under such conditions. Nevertheless, theories of creativity pay only a rather modest interest in the faculty of intuition; it remains outside the rationalist realm of thinking, operating to fit together bits and pieces of heterogeneous materials to forge an image of what is in a process of becoming. The emphasis on scientific procedures in terms of instrumental thinking has underrated the importance for constructing creative images of the real, the object of investigation in the broadest sense of the term. Hence, the interest in intuitive thinking.

Before we move on to the discussion on intuition, the critique on instrumental rationalities will be addressed.

Creativity beyond instrumental rationalities

One central issue in research on creativity in organizations is the ability to clearly define ideas and thought that are considered creative or which have proven to be creative. The dominant approach is to treat creativity in a functionalist and instrumental manner, that is, to conceive of creativity as something that occurs or happens during certain conditions that can be arranged or managed. This rationalist view has been the dominant perspective in contemporary management theory. Gephart (1996: 95–6) writes:

> Rationality has been the driving force of modern management. Rationality has begun to dissipate in postmodernism, to become a cacophony of local rationalities, but we need to decentre rationality, not abandon rationality. We need to place rationality alongside other human faculties – passion, love, hope, and intuition – in our effort to understand and shape the future of management and history.

Although instrumental rationality remains as one of the main ingredients in management practice and theory, it is important to be open to alternative perspectives. Management is not simply the application of several rational principles, such as those suggested by Frederick W. Taylor; it also draws on 'passion, love, hope, and intuition'. For example, creative work is based on highly technical and specialized knowledge within a particular field, but it is simultaneously dependent on commitment, communication, and experimental thinking. Jeffcut (2000: 125) argues:

> [T]he creative process is sustained by inspiration and informed by talent, vitality and commitment (i.e., a need to create rather than to consume): this makes creative work volatile, dynamic and risk-taking, shaped by important tacit skills (or expertise) that are frequently submerged (even mystified) within domains of endeavor. Hence, the crucial relationship between creativity and innovation (i.e., the process of development of original ideas toward their realization/consumption) remains unruly and poorly understood.

Creative work is never solely the outcome of the instrumental application of a set of management principles; rather, it must always be open to what Gephart calls 'other human faculties' such as passion or intuition. Here, one can point to some of the shared demands on the creative person and what Max Weber (1948) called the scientist in his essay (originally a lecture at the University of Munich) *Wissenschaft als Beruf*, 'Science as Vocation'. Here, Weber argues that scientific work needs to be based on the tragic insights of the scientist that his or her work is always ephemeral and doomed to be outmoded and surpassed in the face of the progress of the scientific discipline: 'Scientific work is chained to the course of progress' (Weber, 1951: 137). In addition, the scientist also needs to realize that the only progress possible is fragmented and occurs in a piecemeal fashion: 'Only by strict specialization can the scientific worker become fully conscious, for once and perhaps never again in his lifetime, that he has achieved something that will endure. A really definitive and good accomplishment is today always a specialized accomplishment' (Weber, 1951: 135). These two existential predicaments need to be acknowledged by the scientist. Rabinow (2003: 99), commenting on Weber's essay, writes:

> Science is not wisdom, science is specialized knowledge. A number of important consequences follow from this situation. First, 'scientific

work is chained to the course of progress'. All scientists knows that, by definition and in part of their own efforts, their work is destined to be outdated. Every scientific achievement opens up new questions. One might say that a successful scientist can only hope that his or her work will be productively and fruitfully outmoded rather than merely forgotten. Second, the knowledge worker must live with the realization that not only are the specialized advances the only ones possible but that even small accretions require massive dedication to produce. Dedication or enthusiasm alone, however, are not sufficient to produce good science, nor does hard work guarantee success . . . The calling for science thus must include a sense of passionate commitment, combined with methodical labor and a kind of almost mystical passivity or openness. The scientific self must be resolutely willful and persistent, yet permeable. Androgynous, if you will.

When creativity is examined, it is important to keep in mind that this 'calling' for research is heavily indebted to human faculties such as emotionality and passion. All science-based innovations are the result of hard work and long-term commitment within a specific field of interest. Consequently, creativity is not the release of the untrammelled creative capacities of individuals, but is rather to be regarded as well-organized and detailed processes monitored by peers and experts. Still, notwithstanding the technological apparatus and scientific procedures that are integral parts of scientific work, creativity is always dependent on passion and curiosity. Therefore, creativity can never be reduced to the level of mere machinery and technologies.

New drug development is based on formal management procedures and on factors that remain somewhat tacit: creative solutions to practical problems, unexpected applications of taken-for-granted knowledge, novel forms of thinking, and so forth (Dorabje, Lumley and Cartwright, 1998). These various minor innovations and procedures draw on what we here refer to as intuition – captured by the metaphor *the ability to see that the dots constitute a line rather than being isolated points*. While what Gephart calls rational knowledge is widely known (and not contested facts, the dots), intuition is oriented towards what is not well known and has achieved the status of 'fact'. In other words, intuition is between the well-known facts and procedures in scientific discovery. Intuition facilitates the ability to apply scientific knowledge and to see consequences of various experiments before formal proof is acquired. Therefore intuition is very important for creativity in new drug development. Studies of scientific work (e.g., Knorr Cetina, 1999;

Pickering, 1995; Lynch, 1985; Latour and Woolgar, 1979) show that scientific work is never as linear, homogeneous, and one-dimensional as one may believe. Instead, controversies, alternative explanations, empirical inconsistencies, and local interpretations always characterize production of 'scientific facts'. In short, a certain degree of heterogeneity exists within scientific knowledge. Therefore, scientific and laboratory work is never the 'black box' it is considered to be in common sense thinking. Instead, intuition – the ability to anticipate results and to see broader pictures on the basis of empirical observations – is a highly useful skill. For instance, Lynch (1985), studying work in biology laboratory, observed that laboratory scientists were just as concenred with absences as what what could be actually observed on the microscope photographs:

> [T]he most interesting (and problematic) artifacts were not the definite 'things' that arose as part of specific (and often controversial) accounts. As possibilities they were not, as yet, specific features of any microscopic scene, but were tied to readings of the scene. Such possibilities were often mentioned as *absences* in an observation rather than definite constructive processes (spots, blotches, blurs in a photograph which can be seen as 'intrusions'). The failure of an expected phenomenon to appear was of interest here for the way in which the absence could be formulated under different conditions as an artifactual absence or as a 'real' absence. Under actual absence research conditions, such absences were troublesome since they were necessarily definitive of any real worldly absences, but could be taken as 'failures' in the technical ways of making a phenomenon appear. (Lynch, 1985: 86)

For a biologist, the ability to understand the 'absences', of necessity in-between what was captured by the microscope, was of great importance. Thinking outside of the mere actualities is then a central skill, Lynch (1985) suggests, of the experienced laboratory scientist.

A series of interviews in a pre-clinical organization in AstraZeneca suggests that intuition is a very important factor in new drug development. The new drug development process requires standard operating procedures and routine work, but benefits from additional creative and inventive thinking. In the analysis, we make use of the notion of intuition developed by the French philosopher Henri Bergson. For Bergson, intuition is a key human faculty, capable of 'thinking movement' rather than 'thinking solids'. While concepts and well-known facts are always appearing as fixed points and positions, the faculty of intuition is the ability to think about change and movement between such points. So for Bergson, intuition is part of all sophisticated, creative thinking.

The notion of intuition

This section of the chapter examines the notion of intuition developed by Henri Bergson, one of the most important philosophers of the twentieth century. During his lifetime, Bergson was very influential in the fields of philosophy, politics and, art. Following his death, Bergsonism became increasingly unfashionable and was essentially abandoned until the end of the twentieth century. The start of the new millennium, there was a revival of interest in Bergson's philosophy (see, *inter alia*, Linstead, 2002; Wood, 2002). To discuss Bergson's view of intuition, one must briefly recapitulate other areas of Bergson's thinking. In this way the notion of intuition is placed within a broader ontological and epistemological framework that gives sense and meaning to the notion of intuition. For Bergson, the basic ontological principle is that the world consists of processes. Processes and movements constitute the world that we can experience – not entities:

> In reality, things are events of a special kind, temporary crystallization of images; it would be proper to say that, for Bergson, movement is the real and original stuff the world is made of, whereas the picture of the universe as consisting of distinct material objects is an artifact of intelligence. These ideas – the logical and metaphysical priority of events over objects – was to be subsequently taken up and developed in detail by A.N. Whitehead (1968), probably not without inspiration from Bergson. (Kolakowski, 1985: 45)

This ontological principle is also an epistemological principle. Being in the world is not based on a series of succeeding points, but is instead based on what Bergson calls *durée* (duration). Moore (1996: 55) explains:

> It is not that we start from discrete items of experience spread out in time but somehow threaded together like beads on a string of consciousness. Rather we start from the experience of temporal flow. Temporal structure is not a matter of putting together given discrete items. On the contrary, so-called discrete elements are only apparent when we have a need to pluck them from our *continuing* experience.

Linstead (2002: 101) further clarifies:

> Bergson argues that human experience of real life is not a succession of clearly demarcated conscious states, progressing along some imaginary line (from sorrow to happiness, for example) but rather a continuous flow in which these states interpenetrate and are often unclear, being capable of sustaining multiple perspectives.

Human beings do not experience time as a mechanical stepwise movement from the past to the present and into the future. Instead, they experience time as a continuous series of events based on simultaneity – past, present, and future are never entirely separated; instead, they are always related in experience. Massumi (2002: 200) writes:

> The basic insight of Henri Bergson's philosophy...is that past and future are not just strung-out punctual presents. They are continuous dimensions contemporaneous to every present – which is by nature a smudged becoming, not a point state...Past and future are in direct, topological proximity with each other, operatively joined in a continuity of mutual folding.

In summary, Bergson's notion of *durée* refuses to treat human experience as a mechanical, spatialized experience that consists of clearly demarcated solids brought together. The *durée* of a human being is always recalling the past into the future to anticipate the future.

While the notion of *durée* is primarily a construct that explores the psychology of humans, the idea of continuity and what Bergson calls spatialized thinking is very important for his theory of knowledge. Just as what Bergson calls mechanical clock time tends to break down the continuous experience of *durée* into isolated points and positions, concepts and representations perform the same operation for knowledge. Language, the primary medium for thinking and knowledge, is based on concepts that are generally thought of as denoting certain events, essences, or practices. For Bergson, concepts can only capture a subset of human knowledge because they represent what he calls 'cinematographic thinking', that is, snapshots of events and occurrences in a continuous intrinsically moving reality. Concepts are thus formed as attempts to glue a world in motion into certain positions and fixed points. Bergson (1919: 137) writes: 'To know a reality in the ordinary meaning of the word 'to know' is to take ready-made concepts, apportion them, and combine them until one obtains a practical equivalent of the real.' He continues:

> Every language, whether elaborated or crude, leaves many more things to be understood than it is able to express. Essentially discontinuous, since it proceeds by juxtaposing words, speech can only indicate by a few guideposts placed here and there the chief stages in the moment of thought. (Bergson, 1912: 125)

Concepts are thus ready-mades that are applied to cases; they represent 'classified thinking' (Bachelard, 1964: 75) and are therefore incapable of conceiving of movement. In brief, concepts are 'solids': 'According to his [Bergson's] account, concepts are formed on the model of spatial solids, and it is consequently impossible to think about time without importing into it some of the features of homogeneous space' (Mullarkey, 1999: 19).

Thinking always uses concepts, and concepts can never entirely capture the movement and becoming of being, although they are still useful tools in understanding such a world. For Bergson, intuition enables us to understand movement. While concepts consist of solids based on cinematographic thinking which make us unable to see what is outside of ourselves, intuition is the faculty of thinking in-between the solids. Grosz (2001: 175) explains:

> Intuition is our nonpragmatic, noneffective, nonexpedient relation to the world, the capacity we have to live in the world of excess of our needs, and in excess of the self-presentation or immanence of materiality, to collapse ourselves, as things, back into the world.

Ansell Pearson (2002: 124) adds:

> According to Bergson, the abstract intellect, which has evolved as an organ of utility and calculability, proceeds by beginning with the immobile and simply reconstructs movements with juxtaposed immobilities. By contrast, intuition, as he conceives it, starts from movement and sees in immobility only a snapshot taken by our mind.

In other words, intuition is what breaks free from language and sees what is outside of the concept, outside the language that we use to denote the world. Here, language is not only everyday concept, but is equally the regime of representation that dominates the world of natural sciences and the biosciences. While language is a prosthesis for thinking – just a tool – thinking that draws on intuition abandons such a prosthesis in favour of a more free form of thinking. In Whitehead's (1938: 49) formulation: 'Language halts behind intuition.'

In summary, Bergson develops an ontology and epistemology of movement and becoming; processes rather than solids and entities constitute the world. Therefore, the human experience does not consist of single instances stacked one on top of another but is based on the simultaneity of past, present, and future. In addition, Bergson's theory of knowledge separates use of ready-made concepts that are serving as tools for thinking and communication and calls that which is positioned between concepts (solids) *intuition*. While concepts help us see the world as a series of demarcated instances and events, intuition makes us think in terms of movement and becoming. Intuition is thinking that lies in-between the known and the represented; Intuition is thinking beyond language.

In terms of creativity, and, more specifically, creativity in terms of new drug development in the pharmaceutical industry, intuition is thinking that uses what is already known – solids of verified knowledge or facts provided by the research efforts and laboratory work – to anticipate what is not known and established, negotiated, and agreed on as facts. Intuition is thus thinking that goes beyond or passes what is already known to enable new solutions and findings. As the empirical material suggests, this form of thinking outside of the solids is highly valued in pharmaceutical research.

Intuition in new drug development

Our study investigated different aspects of intuition and its relation to organizational creativity in pharmaceutical research. Pharmaceutical research in the discovery organization is based on sophisticated techno-scientific laboratory work. The search for NCEs and their further development to finished products occur in distributed knowledge systems in which several different areas of expertise are integrated. NCEs are the outcomes of joint efforts by medicinal chemists, biologists, and pharmacologists. Questions and issues addressed in the study were divided into three categories. The first defined and positioned phenomena in the context of pharmaceutical research. The second investigated how intuition plays a role in drug discovery research and also examines its relation to organizational creativity. The third dealt with different organizational factors, such as technoscience and leadership and their relationships to intuition. For example, how different sophisticated technologies and technoscience influence interaction with phenomena.

What is intuition in the context of pharmaceutical research?

Because intuition has multiple connotations, it is important to specify its meaning in the context of pharmaceutical research. Policastro (1995) suggests two complimentary definitions of intuition: one based on a metaphorical perception of phenomena and the other on a tacit form of knowledge. One of the respondents expressed this latter form and emphasized that intuition in pharmaceutical research is a combination of broad knowledge and competence:

> Intuition comes from broad competence together with extensive experience in a special area. For me, intuition is the ability to predict things with pretty good precision on the basis of the competence platform somewhere in the background...yes, it's like a limit between intuition and not yet proven knowledge is floating as a chemist. I mean, you can show a chemist a structure and say: 'Do you think that this will be potent?' And then he has a much better opportunity to answer yes or no to the question than another chemist who has not worked in our project. He can't point it out because that nitrogen is there or there. He can look at the structure and say: 'No, I don't think so.' It is probably doubtful. It is obvious he can point out certain things, like: 'I think the chain is a little too long.' Or something like that. (AstraZeneca, Rearcher, Discovery, Pharmacology)

Another respondent defined intuition as a feeling and expressed a significant amount of vagueness and thus correlated intuition with risk taking:

> I think that intuition is a kind of feeling. It is like what vision is for planning. Intuition is a type of capacity that comes with experience. I actually think that intuition is very important in the whole research process – particularly in early discovery phases. Because intuition is correlated with risk taking, it's difficult to base decisions on intuition for clinical programs, which is of course ethically correct. But in early phases in discovery, it's simpler and easier to take intuitive decisions about different things, like in toxicity studies or choices of methods. (AstraZeneca, Researcher, Discovery, Biochemistry)

One researcher stressed that intuition represents a dichotomy that is: (1) rational and sensible; and (2) irrational and impossible to communicate:

Intuition is something that summarizes experience for better or worse, because sometimes it's rational and other times it's irrational. I believe that many experienced researchers have some kind of touch of recognition to identify new situations and put them in relation to something they have experienced before. I think that often, you see that intuition is something that you cannot put into a list. I believe in this method or in this molecule because...And you can list carefully researched facts about it and maybe refer to different parts, and there you have it very clearly why you recommend one. But many times it is a little weaker. Maybe you have made some calculations that are not very clear, but you have made others and maybe have seen earlier cases that remind you of it, but you cannot really put your finger on what it is. Like when it's more vague but you anyway feel very strongly that this is the one you believe in. (AstraZeneca, Senior Researcher, Discovery, Computational Chemistry)

An organic chemist described intuition as the ability to make combinations and explain how chemical structures can be visualized into a type of harmony:

For me, intuition is almost emotional; it's like that things look good. For example, if I have a synthesis that I am working with, I can get a feeling that 'this should work'...it's something that is very useful in what I do, because it has to do with combining earlier pieces of evidence – call it intuition. But it is imagination and an ability to make combinations. It's like you feel intuitively that it's right. (AstraZeneca, researcher, Discovery, Medicinal Chemistry)

One respondent emphasized the strong relationship to knowledge and pointed out the ambiguity of intuition in the research process and how to handle it:

Intuition is based partly on the experience of having being a part of and seen many examples and then being able to connect the experience with...But also being able to digest many different signals into a conclusion – that's some kind of partial explanation of what intuition is, I think. The difficulty is that if you look at calculation methods, intuition is like neural networks; there is no explanation. You can train yourself in calculation models, calculating responses that you see are pretty good, but you do not have the vaguest idea of how the algorithms have come to that result. But there are other robust calculation methods that are more rational, where you can understand the coefficient that appears. And man's brain has an ability to weigh in all types of information and backgrounds and experiences into something that is a decision or an intuition or whatever you can call it. And it is good when it is rational. It is less good when it is irrational. I think we are coloured by many irrational things. (AstraZeneca, Researcher, Discovery, Computational Chemistry)

In conclusion, all respondents expressed to varying degrees that intuition in the context of pharmaceutical research is an intinsic ability to produce various associations for which experience and broad knowledge play an important role that may lead to important solutions for scientific problems.

Does intuition matter in drug discovery research?

On the rather broad issue of whether intuition plays a role in drug discovery research, all respondents argued that intuitition has a major, but complex influence. One respondent explained:

Of course, it's very important. But I firmly believe that intuition is a summarized picture you get from all the experience and the knowledge you have. So I don't think that intuition is hocus pocus or something that you should be sceptical about. It's a gut feeling. Very important. And I think that it's based on things that are inside you and that you should absolutely trust it. (AstraZeneca, Senior Researcher, Discovery, Computational Chemistry)

All respondents expressed in various ways that there is a relation between intuition and creativity. Many respondents claimed that intuition and creativity overlap considerably. In some cases, two respondents thought that intuition and creativity are essentially the same concept. But they emphasized that intuition cannot be controlled and creativity can, for example, through imagination and domain knowledge: 'You cannot do something if you don't know the tools: the carpenter must know his tools.' One researcher provides a concrete example of how intuition links to creativity:

Intuition and creativity go together. It's not so easy to separate them. You also must have intuition; it's not always that it must be that way. For example, entrepreneurship need not always lead to getting there the quickest way. There can be other people who help make a decision. It's exactly the same thing that you do in the lab. You might have a target molecule but a lot of different ways of getting to it. Not only using intuition but a combination of intuition and experience, in any case; one person maybe chooses a way that leads to being able to make the substance more quickly than the others, for example. That you really can produce it. (AstraZeneca, Researcher, Discovery, Medicinal Chemistry)

Another respondent gave this example, which points out that intuition can also be an obstacle to creativity:

Yes, the connection is probably complicated. I'd say that creativity can be damaged by too much intuition and especially this unconscious intuition. Then I'm worried about getting stuck in a rut. That in some way, it may be wrong to say that it quantitatively obstructs creativity, but it stops it, I think there is a risk that it stops creativity qualitatively in a narrower niche. If you dare to challenge and question your own intuitive solution to problems, then you maybe broaden your perspective and come to – I would not say more or less creative, I mean not fewer or more creative solutions – but you may reach other qualities, and they in turn are difficult to evaluate, which is better or which is worse? (AstraZeneca, Researcher, Discovery, Computational Chemistry)

Although the notion of intuition is perceived as something important, almost all respondents said that intuition is rarely, if ever, discussed in the organization:

It [creativity] is discussed at times. I am one of those people who makes just these, if you could do rational methods for working instead, uses experimental design, thinks through why you do an experiment and so on. Many people say: 'Yes, when we're designing drugs, we must let the chemists use their intuition, you know?' And that makes them say: 'Yes, but I've darn well been working with synthetic chemistry for 20 years, and I get a feeling that if we put an amide group here, then there will be higher activity'. And then the discussion comes directly into: 'Yes, to what extent should we let the chemists use their intuition?' And of course, it does happen that structures that you find out are good have been intuitively designed.

And then you can ask yourself if it's what we call serendipity and how much serendipity is colored by intuition that someone has through experience. But research can be maximizing the chance to have serendipity, although where intuition is maybe a positive factor. Sometimes anyway. I think it's very important to be aware of intuition. Going on intuition without knowing it yourself. It's like analogous to being unaware. Lack of knowledge is pretty safe, but not knowing about your own lack of knowledge, unawareness, that is not good. And I think it's the same with intuition, except in the other direction, because intuition can be a good thing, especially if you understand it and deal with it in a healthy way. (AstraZeneca, Senior Researcher, Discovery, Computational Chemistry)

Here, a researcher contrasted the role that intuition might play (although it's a vague, less controllable concept) with the present organization and its strong emphasis on cost-effectiveness, detailed project plans, and a controlled drug discovery process:

The drug development process is not so damned rational as a lot of people would like it to be, instead intuition can prove to be extremely significant. And intuition comes, I think, you can make it easier for them by having a long-term view in the disease area. I think it's much easier to follow your intuition if you have worked within a specific disease area for a long time than when you have more general intuition about different things. It's like a feeling when you just read a scientific paper you feel that this – sometimes you can have some kind of worried feeling – you feel that this is important for our work, but you don't know what. And then you can't let it go. And sometimes then the whole thing gels and you realize: 'Yes of course!' And then if you have even more luck, it can lead to success. And without what you could define as intuition that gave some person that feeling of worry, like: 'Damn, I can't put this thing aside.' You can't fall asleep at night, and you don't really understand why you have that worried feeling. It must be some kind of intuition that you have. But at the same time it's actually – you can actually describe it like you have broad competence. I think that the people who are good researchers are the ones who have the ability to store information in their heads and bring it out and remember that: 'this doesn't really go together with the article I read seven years ago', and they get it out and look at it. Okay, it must be because of this. That ability to be able to store that information and retrieve it. (AstraZeneca, Project Manager, Discovery, Pharmacology)

A specific example of how intuition plays a role in drug discovery research is given in the following example. The researcher, in organic chemistry and computational drug design, was involved in a screening project that was looking for a new cardiovascular drug compound. The task was to invent a pathway for how to synthesize a new chemical structure. Computational chemistry is based on computational technology being able to visualize and simulate the way in which drug molecules and targets (e.g., large proteins or an enzyme) may interact.

> [I]t can be small things, you might get something back – you work in a project, you get back data and get the feeling that something is wrong. It's just not right. You look at the pattern and see for example which structures are active, which are not active, you look at the pattern and you feel that 'no, this is not right.' And what do you do then? Of course you try, for example, you screen them again, you test them one more time and find out that it is wrong, everything does not really work the way it should or that it really is what it looks to be, and then there's still something that is not right and then you must go further and then maybe it has something to do with the mechanism. You must keep on working and modifying. You have to maybe get to the bottom of the thing that is not right. The picture is not completely clear. And you proceed in that way and discover something else. Yes, you get the feeling that this is not quite right. And you work on it, make sure that you go on trying to get to the bottom of it. I have had that experience in projects. (AstraZeneca, Senior Researcher, Discovery, Computational Chemistry)

In conclusion, all respondents express the notion that intuition in different perspectives and disciplines plays an important role in drug discovery research. In addition, most of the respondents argued that intuition is strongly linked to creativity.

Intuition and organizational aspects

To make the research more effective and increase innovative output, the influence of various technologies (e.g. high-throughput screening and computational drug design) is now an important part of pharmaceutical research. These new scientific screening methods represent an attempt to use various forms of what Bachelard (1984) calls *phénoménotechniques*, 'technologies of visualization' that enable the faster identification of NCEs. The technologies involve a completely new way to manage scientific data and information. These technologies might make routine work more efficient but may be an obstacle to scientific creativity (see Thomke and Kuemmerle, 2002: 631; Cardinal, 2001). However, as a representative for top management pointed out, the increasing role of intuition might have to bridge that gap:

> [I]f you can use technologies in routine work and make things more quickly and perhaps more precise, get more reliable results like a lot of these robot systems can do than if you are doing it manually then it is a great advantage, which should also actually give people more time to think creatively. Many successes in automatization make it possible for us to have access to completely new amounts of data that we can treat in a completely different way than before, because we have so much more data. We can see patterns and other things that maybe would otherwise be completely impossible to identify. You should look at these technologies as tools and then there is always a human factor when you are looking at data. This evaluation, you have to put it into its context. Is it actually reasonable? Should we choose this chemical structure that had a signal in this high throughput screening? Or is it perhaps completely impossible to do, modifying it so that it can be optimised. We don't have a really good selection system yet. Instead, we have certain filters so that you can take away characteristics and other things but it's still up to a creative evaluation by an experienced chemist, and it's also based on – not just knowledge but also on intuition. I think that intuition and research are incredibly important when it comes to seeing that this is darn important. It may also be creativity. It's a question of definition, you know, but having an intuitive feeling about this being the right way and this is an important result. (AstraZeneca, Senior Management, Discovery)

Most respondents disclosed problematic factors when dealing with intuition in the organization. One factor was the way in which intuition is increasingly affected when planning and managing drug discovery research. This is illustrated by one respondent:

> The organization today is in such a streamlined format in some way, it feels like. The way it works with us anyway, you maybe work for a period of months with a target and then that is the end of that and you start on something new and it can in and of itself be very stimulating, but after that time you have learned what you have started to work with and then you must stop and start on something new.

I don't know – in that way I have a little bit of a hard time thinking that we're working in the right way somehow. There must be continuity in some way in the organization and it is a little too divided into parts in some way. Yes, I think that it actually feels like that sometimes. I hope that there is room for continuity too, but they have not done that anyway. I mean, the press gets harder and harder on the organization too. There must be more and more targets, there are projects and so on and so forth. I mean, there is hardly time for being able to sit down and think. You must put together reports for different levels all the time. You are driven by having to have something positive to say at these meetings, and you focus on coming up with something for them, but that maybe is not really what you should be doing after all, but maybe you should work a little, little more long term, and you miss that with this type of project. (AstraZeneca, Senior Management, Discovery, Organic Chemistry)

Another aspect of intuition is that it is seen as something mysterious, and subsequently unprofessional or non-scientific; a chemist explains:

There are prejudices. I mean that intuition is built on – like I said, what I believe – earlier experience. And you can easily be led to believe something that is a preconceived idea and that directs you too much, and you don't look at the facts that exist. (AstraZeneca, researcher, Discovery, Computational Chemistry)

The importance of leadership and intuition may not be obvious. But a representative from senior management pointed out the need for management attention in relation to intuitive dimensions in drug discovery:

Yes, because as a leader, the point is not only to push your own ideas, you know, but to listen to the ideas of the person who is the most recent employee. I think it would also send the right signals if even higher management can accept, so to speak, the newest guy's view of the business. As it looks through others' eyes, too. They come from the outside. (AstraZeneca, Senior Management, Discovery)

In conclusion, although intuition in drug discovery research is claimed to be highly important, the respondents argue that it is something that is only occasionally or never talked about in the organization. And the common notion from the respondents reflects different concerns about how intuition is exploited in the current rationalized drug discovery process.

Modes of thinking in creative work

Pharmaceutical researchers claim that the human faculty of intuition represents thinking that goes beyond the strictly rational and representational; intuition is claimed to be emotional, a 'gut feeling', drawing on experience, and oscillating between being rational and irrational. Intuition is a mode of thinking that accounts for what is not really proved in scientific terms, but is still nevertheless valuable knowledge in the process of drug discovery and development. In addition, the faculty of intuition matters in new drug development. Intuition is a type of thinking that is captured by metaphors such as *thinking outside the box* or *seeing the broad picture* – that is, metaphors that depict intuition as the ability to see relationships, causalities, and other associations when there are not yet proofs of such relationships. New drug development is a highly specialized activity that consists of many different scientific disciplines. And authorities regulate the process. Consequently, the effective management of operations must support new drug development. If intuition is regarded as the capacity to make decisions under time pressure – without complete information (i.e., being subject to what Herbert Simon calls *bounded rationality*), then intuition is a highly useful resource in new drug development. But intuition is, as some of the interviewees pointed out, by no means an extra-rational or super-rational capacity that can be invoked in cases for which complete information is unavailable; intuition is always at stake because it draws on experiences and emotional faculties. So invoking intuition is a political issue because, by definition, it goes beyond formal decision-making systems that are provided. In short, intuition is an individual and organizational resource that is difficult to manage and unlike new technologies, such as HTS, which do not rely on the experiences and emotions of pharmaceutical researchers but rather on the automation of the identification of new chemical substances.

The pharmaceutical researchers emphasized that intuition is an important resource in new drug development activities. Yet the concept of intuition is not fully examined in the organizational creativity literature. If one follows Bergson in conceiving of intuition as thinking that is 'pre-representational' and operates outside the favoured regime of representation (e.g., mathematics or a scientific vocabulary), then intuition has a rather clear meaning and role vis-à-vis more conventional and analytical forms of rationality. For Bergson, rational thinking is *analytical* in terms of being able to reduce a complex matter to a signifying system, but intuition is *synthetic* in terms of being able to see what is outside of the signifying system. When using the *single-dots-constituting-a-line* metaphor, then rational thinking is the individual dots (which we know are there) while intuition enables synthesis from the line and substantiates the claim that the spaces between the dots are not just voids, but are regularities that constitute the line. Consequently, it may be argued that the literature on organizational creativity has not been particularly concerned with the pre-representational forms of thinking represented by intuition. It is common to address extraordinary contributions and individuals in this literature, but there is no coherent theoretical framework developed for use when studying such events and occurrences. Therefore, a Bergsonian view of intuition could be fruitfully developed within this literature. Rather than conceiving of some forms of thinking as being merely 'original' (one trait of what we tend to deem as creative), Bergson's thinking offers an ontological and epistemological model that can examine what this kind of originality consists of, for example, if originality in solutions will make interesting and new syntheses of what is already known. At the bottom line, Bergson may be a useful ally when criticizing the technical-instrumental rationality that serves as the bedrock for all management activities. Standardized management solutions for engagement with an external world (e.g., calculation, reduction of continuous realities to discrete events and entities, and enactment of stable and predictable relationships between different actors) are mostly analytical in nature; the management ideology conceives of a world that is manageable (O'Shea, 2002: 123–4; Gephart, 1996). This works fine as long as such reductionism is applicable. In many cases, management practice cannot rely on its analytical apparatus and needs to develop practices, techniques, and systems that can deal with fluidity, movement, and change, in brief, when speaking of Bergson, what cannot be fully captured by the rational thinking of the intellect. Grint (1997: 9) argues:

> Like many other forms of thought, [management theory] does tend to rationalize away the paradoxes, chance, luck, errors, subjectivities, accidents, and sheer indeterminacy of life through a prism of apparent control and rationality.

In the case of new drug development, perhaps what occurs between analytical systems of HTS and other technologies are never regarded as anything more than such chance, luck, and errors. Thus intuition may be representative of a form of thinking that goes beyond these reductionist modes of thinking. Arvid Carlson, the 2001 Nobel laureate in medicine, and a man with extensive new drug development experience, testifies to this need for taking the consequences of what one may already know:

> Especially on the discovery side, it is like walking in a labyrinth, you face many decision points and the thing is not to jump in the wrong direction too many times. The first thing you need is luck, and then it is the other, what people call intuition...And then there is the question: what is intuition? Intuition is probably just that, of having a very incomplete, a very fragmentary basis and of being able despite only having fragments to see a pattern that leads your decision in a certain direction. (Arvid Carlsson, Interview, 2003)

Being able to theorize the fragmented, incomplete world inhabited by pharmaceutical researchers remains a challenge for the organizational creativity literature.

Implications for management

Implementation of a more rationalistic approach to become more effective has been the dominant trend in many large R&D organizations. Many pharmaceutical companies have turned to rigorous project and portfolio management in order to make research more efficient (Schmid and Smith, 2002b). One could argue against the trend to implement policy that it is too rigorous. This study suggests that intuition is an intrinsic part of the creative process in drug discovery and thus an important organizational resource (Sundgren and Styhre, 2003a). The study's narratives suggest that intuition and creativity are poorly institutionalized in research-based organizations. As a consequence, rationalist approaches that draw on technoscientific practices (e.g., HTS) would benefit from being supported by continuous, widely shared narratives on how research, innovation, and creativity materialize in daily activities in pharmaceutical and other research-based organizations. The narrative view of organizational practices (see, e.g., Czarniawska, 1998; Gabriel, 2000) suggests that the process of organizing is embedded in storytelling and joint sense-making of events and occurrences. Narrative studies of organizations, such as Orr's (1996) study of copy-machine technicians, Boje's (1991) study of an office supply firm, Bryman's (2000) examination of technology-based firms, and Humphreys, Michael and Brown's (2002) analysis of organizational identities from a narrative perspective suggest that the institutionalization of vocabularies, standard plots, speech genres, and so forth, support and reinforce organizational practices. This study suggests that an ongoing narrative on intuition and creativity would facilitate more effective research practices. Being able to tell stories and share experiences from highly specialized, sophisticated research-based work remains as one of the key mechanisms that underlies excellent organizational performance in the pharmaceutical industry. Thus, telling stories about intuition and creativity is an integral, yet somewhat neglected component of the pharmaceutical researcher's set of skills.

From the rationalist viewpoint within contemporary management theory and from a pharmaceutical industry perspective, the rationalist view of making research more effective has some parallels with the dominant knowledge management tradition (Styhre, 2003). This tradition tends to manage and distribute knowledge in organizations as fixed and ready-made. This trend to codify, integrate, and reproduce knowledge in the pharmaceutical industry offers only limited latitude for creating new knowledge. And this could be why the industry lacks radical innovation (Horrobin, 2002). We argue that the role of intuition is an important subset of understanding organizational creativity and a rather unexploited platform for creating new knowledge, which demands receptiveness to a more critical view of traditional knowledge management theory.

Creativity in the context of the pharmaceutical industry is an ambiguous concept (Sundgren and Styhre, 2003b). The predominant notion of creativity stresses something that is purposeful; something that other scientists have not done before. Creativity must always be based on an accurate knowledge of the specific domain. Thus, organizational creativity in new drug development demands an organizational capacity for becoming masters of a specific scientific domain, while allowing for an overview of an area of science.

A clear message for senior management is to be open to discussion within the organization regarding ways in which intuition plays an important role within drug discovery research. This would enable a better understanding of organizational creativity in which intuition not is seen as a fuzzy concept but, rather, as an asset that could be in balance with the rationalistic thinking.

We argue that intuition is needed, because creativity is an ambiguous concept. As Deleuze and Guattari (1995: 18) write:

In any concept there are usually bits or components that come from other concepts, which corresponded to other problems and presuposed other planes. This is inevitable because each concept carries out new cutting-out, takes on new contours, and must be reactivated or recut. On the other hand concepts also has a becoming that involves its relationships with concepts situated on the same plane.

Acknowledgement of intuition's role in new drug development would: (i) enrich contextual thinking that broadens scope through radical thinking and enrich the concept of organizational creativity; (ii) increase an organization's ability to move between different scientific domains within new drug development; and (iii) enable management to increase the probability of capturing ideas in an early phase, which could result in scientific breakthroughs.

Conclusions

This chapter has presented a study of the role of intuition in pharmaceutical research. Conceiving of intuition as being the ability to synthesize on the basis of available information, intuition is contrasted with reductionistic and analytical forms of thinking. Even though the interviewees argued that intuition is a highly useful human faculty, it is still somewhat controversial to use intuition as the basis for decisions. Because intuition was more closely associated with emotionality and embodiment (for example, as captured by the proverbial expression 'gut feelings') than with cognitive capacities, intuition served as some kind of supplement to more conventional forms of thinking. Still, intuition is acknowledged among pharmaceutical researchers, especially in cases in which the researchers must account for a multiplicity of facts during decision making. Consequently, the organizational creativity literature that is highly positive to the idea of the extraordinary and ground-breaking innovations, must recognize the faculty of intuition. This chapter suggests that the thinking of Henri Bergson may be one fruitful resource to exploit in this endeavour. From an innovation management research policy perspective, this study also calls for more critical perspectives in knowledge management theory regarding how to understand and create new prerequisites for the creation of new knowledge.

7
Leadership and Creativity

Introduction

The leadership, or management, angle is nearly always absent from the literature on organizational creativity (e.g. Mumford *et al.*, 2002; Jung, 2001). There are several reasons for not taking leadership into account in traditional creativity research. The majority of research has focused on distinct aspects of creativity, among them strategy (Parnell *et al.*, 2000), structure (Damanpour, 1998), climate (Ekvall and Ryhammar, 1999), individual performance (Runco and Sakamoto, 1999), group performance (Amabile and Gryskiewicz, 1989), and dissemination practices (Abrahamson, 1991). However, as Mumford *et al.* (2002) argue, management is conspicuously absent from the list of potential influences. Management, and leadership, at least traditionally, has not been held to be a particularly significant influence on creativity and innovation. According to several scholars, one reason is that we tend to discount leader influences and creativity may be found in our *romantic conception of the creative act* – a conception where ideas and innovation are attributed to the heroic efforts of the individual. One aspect that discounts leadership and creativity may also be that the professionalism, expertise, and autonomy that seem to characterize creative people act to *neutralize*, or *substitute* for, leadership (Mumford *et al.*, 2002).

If one adopts a system perspective when considering creativity in organizations, then leadership and management influence become central issues. Furthermore, it is management that decides what is creative or not, and makes a decision about 'how much creativity' they believe satisfies the need for the organization to renew its product or service portfolio. Thus, it is easy to creative action in organization is intertwined with management

The investment in creative activities will always produce certain types of organizational activities – the consequences of which may not always be desirable. Moreover, by definition organizational creativity implies a deviation in some sense from the standardized way of doing things, which includes, for example, factors such as persistence, flexibility, and opposition. Project groups may reformulate problems and objectives when facing problems, rather than continuing down the same path. Because creative processes are always non-linear and disruptive, and are based on the interaction of tight and loose systems, creativity is costly and places great demands on resources that must be managed and controlled. Although creativity is not a good thing per se, it can also be detrimental to organizational activities in cases where stability, predictability, and manageability are highly needed and praised. Creative activities are at the very heart of organizational renewal, but it may be that it is misplaced at times. In summary: organizational creativity, in Whitehead's (1927) formulation, begets the new. As such, it always challenges the existing culture and power structures (Staw, 1995); structures that are difficult to change.

This chapter is structured as follows: First, the notion of management and leadership is discussed. Secondly, a brief theoretical overview of the management and leadership that is relevant to influence organizational creativity is discussed. Thirdly, some examples of management practices from four companies – AstraZeneca, ACADIA, Carlsson Research and Wingårdhs Architect Firm – each of which has demonstrated high levels of organizational creativity. In various ways these leadership behaviours contribute to creativity and innovation in an organizational setting. These practices represent different examples of management abilities to handle paradoxes. Next, the *Creative Equilibrium Model* built on the capability to handle the presence of both 'stabilizers' and 'de-stabilizers', which serves to deal various paradoxes concerning organizational creativity, is presented. Finally some practical and theoretical implications are outlined.

The notions of management and leadership

Managers versus leaders

One important distinction in organization theory is that between *management* and *leadership*. It is not easy to encapsulate the term management in one, single, unified definition. Here we employ the concept of *management*[31] to capture all sorts of processes that are aimed at administrating, monitoring, controlling, governing, and steering a certain practice or activities in an organization (Styhre, 2003). According to Griseri (2002), management is a hybrid concept comprising different processes, events, and qualities. For example, Zaleznik (1977) proposed that managers vary in terms of their motivation, personal history, thoughts, and behaviours. According to Bass (1990), managers plan, investigate, coordinate, evaluate, supervise, and create goals to maintain the stability of the organization.

The second notion, leadership, is equally complex. For the sake of simplicity, in this discussion we adhere to Fred Fielder's (1968: 8) definition wherein the *leader* is defined as follows: 'The individual in the group given the task of directing and coordinating task-relevant group activities or who, in the absence of a designated leader, carries the primary responsibility for performing these functions in the group'. The leader can thus be formally assigned or operate without such mandate. Katz and Kahn (1966: 334) define *leadership* rather more broadly as 'any act of influence on a matter of organizational relevance'. Research on leadership is a long-standing tradition in organization theory and management studies. The first detailed analysis of how leaders actually spend their work time was conducted by Carlsson (1951), who found that the working day of the leader is disruptive and fragmented and composed of a series of brief encounters, meetings and exchanges with co-workers and stakeholders external to the firm. Mintzberg's (1973) study of leaders was very much modelled on Carlson's research and drew similar conclusions about the disruptive nature of leadership work. Philip Selznick's *Leadership in Administration* (1957) proposed a view of leadership that was closely associated with the institutionalization of certain activities and behaviours in the organization: 'Leadership is a kind of work done to meet the needs of a social, institution' (Selznick, 1957: 22). As a consequence, leadership is embedded in social processes: 'Leadership is not equivalent to office-holding or legal prestige or an authority or decision making . . . understanding leadership requires understanding of a broader social, process' (Selznick, 1957: 24).

---

[31] The French term administration – which most English translations have rendered as *management* – is etymologically derived from the Italian verb *maneggiare*. As Jacques (1996: 88) writes, 'meaning to handle, especially, to handle horses – *Il maneggio*, the managerial employee, emerges'.

This leads Selznick to conclude that 'leadership is dispensable'. He continues: 'The idea developed in this essay is that leadership us not equally necessary in all large-scale organizations, or in any one at all the time, and that it becomes dispensable as the natural process of institutionalization becomes eliminated or controlled' (Selznick, 1957: 25). Rather than treating leadership as the *primus motor* of the organization, Selznick regards leadership as being determined by pre-existing institutions and also the new institutions brought into being through the leader. Leadership is then what is helpful and useful during certain conditions while at other occasions there is little need for leadership.

In the 1960s, Fred Fiedler (1968) developed what he termed the 'contingency model of leadership', suggesting that leadership styles and practices always need to be adapting to changes in the environment. Very much in line with Selznick's (1957) argument, Fielder argued that leadership needs to be both situational and context-specific. Perhaps the most frequent framework in the contemporary research on leadership is the distinction between *transactional* and *transformational* leadership (Bryman, 1996). Transactional leadership denotes the day-to-day management of activities, whereas transformational leadership is the entrepreneurial, forward-directed and essentially dynamic component in leadership work. In the management literature, it is generally the transformational style of leadership that is under investigation and is praised as what is contributing to organization change. In the contemporary research on leadership, there is some interest in the influence of what Max Weber (1992) calls *charisma* – authority based on the ability to influence others through one's personality and charm. In this strand of research, the notion of charisma and its function in the constitution of an entrepreneurial self is tightly knitted together with broader social changes increasingly emphasizing symbolic management and performance-based interaction (Flynn and Staw, 2003; Ball and Carter, 2002; Weierter, 2001; Steyrer, 1998). A closely related field of research is exploring the dramaturgical and embodied aspects of leadership – for instance how leaders are making public performances and making use of various insignia to promote their ideas and make a better appearance (Harvey, 2001; Ropo and Parviainen, 2001). In general, a more critical stream of research on leadership is being formed. First, the so-called critical management studies tradition has examined the underlying ideologies and concealed fallacies in the leadership discourse, with the intention of pointing out the frail epistemological grounds for the portraying of leaders as being superior to their co-workers in terms of know-how, efficiency and ethical standing (Alvesson and Svenningson,

2003; Alvesson, 1992, 1996). In this view, leadership is a social practice that tends to become romanced and veiled by various ideological objectives. The other critical tradition engaging with leadership is the feminist orientation, both pointing at the theoretical underpinnings of leadership (Calás and Smircich, 1991) and conducting empirical studies. What is of particular interest to feminist researchers is the relative underrepresentation of female leaders such as CEOs, board members, professors, and so forth (Dreher, 2003). Some feminist leadership researchers even claim that the very idea of leadership is a masculine construct, refusing to acknowledge any female experiences or competencies:

> Maleness and masculinity are the templates for leadership. Within the confines of technological rationality, leadership has been constructed on the basis of male experience, but this experience has been universalized, and women have been labeled as deficient leaders. By definition, they lack what they can never attain. Men are the norm, women the deviant, the different, the lesser. (Oseen, 1997: 175)

Both the critical management studies tradition of research and the feminist reception of leadership research point to the underlying assumptions and beliefs about leadership and the role it is supposed to play in organizations.

In this literature, *management* in most cases denotes the formal, scientific, and present-oriented process, whereas *leadership* includes the informal, flexible, inspirational, and future-oriented process (e.g., Kotter, 1987). Leaders are often portrayed as visionaries who inspire workers to take part in individual training and competence development and the organizational change projects (Sternberg, 2003). In addition, the discourse on management in innovation and creativity literature is somewhat heterogeneous. Concepts such as leadership and management are often intertwined and lack clear definitions. As a consequence, the research on leadership is not to be treated as a progressive science with clearly established methodologies and a firm theoretical corpus, but is rather a loosely coupled patchwork of theoretical debates, empirical studies, a vast amount of popular business press articles and a great many consultancy and management training services, and, above all, the day-to-day practices in organizations and companies.

Leadership in science-based firms

Managers in new drug development activities in pharmaceutical companies are not easily placed in conventional categories. In practice, leadership is more complicated and ambiguous. Research in the industry suggests an image of the manager not as an administrator, but rather a person whose primary interests are science and research. Traditional management practices and managerial concerns come at best second. In the case of AstraZeneca, this is supported by the fact that line and project managers have high academic credentials – often doctoral degrees in biomedical science. Several studies testify to scepticism and unwillingness to discuss traditional management practice in life science industries (e.g., Llewellyn, 2001; Kitchener, 2000; Parker, 2000). For example, in a study of a health care company, Llewellyn (2001: 604) found such unwillingness to keep managerial issues at arms' length. She writes: 'Any clinician taking up a management position – even with the medical establishment – risks loss of respect and clinical visibility.' So compared to the role of science and research, management concerns have a relatively lower status within the industry (Kalling and Styhre, 2003). Finally, Llewellyn's thesis uses a somewhat extended definition of management in new drug development. Managers control and govern the daily work, but also act as scientists: supporting, guiding, and protecting project ideas. In this sense, management is assumed to be competent to evaluate and decide what is or is not creative in the research process. It is within this environment that organizational creativity in new drug development is discussed. In addition, the empirical section of this chapter draws on a study of leadership in an architecture office. Although we have little evidence based on previous research from the field of architecture work, we suggest that the same reluctance towards formalized leadership practices are possible to observe among architects and other staff at the bureau. This does not suggest that leadership is not of relevance for the performance of the bureau – on the contrary, most interlocutors in the study agreed that is the case – but leadership practices are different from those in other industries. While the science-based innovation activities in the pharmaceutical industry is strongly influenced by a scientific agenda and a passion for scientific work, the work in an architecture bureau is equally dependent upon what may be called a practico-aesthetic ideology wherein both practical civil engineering concerns and considerations and aesthetic visions are mingling and influencing the activities in the course of action. Therefore, the same sceptical attitude towards the mainstream management discourse image of leadership is present in the two different environments. Leadership is not what is imbued in all industries but is at times an additional feature.

Theoretical aspects of leadership and creativity

There is a somewhat diffuse notion of the leadership qualities needed to support creativity. Or as Mumford and Connelly (1999) summarize the research findings on leadership aspects concerning creativity. A number of scholars have examined the kind of environment that leaders should create if there is a need for organizational creativity. Broadly speaking, the result obtained from this research is that less tightly structured, less bureaucratic organizations facilitate creativity. Leaders moreover should try encourage openness to new approaches: focusing on processes as outcomes; permitting autonomy and risk taking; rewarding creativity and innovation; providing demanding, intellectually challenging environments; building feelings of self-efficacy in subordinates. Sternberg (2003) writes concerning creativity in leadership:

> Creativity is important for leadership because it is the component whereby one generates the ideas that others will follow...Many leaders are academically and even practically intelligent, but uncreative; they lead people through their ability to influence rather than through their agenda. (Sternberg, 2003: 391)

In the extant leadership literature, whether it deals with leadership traits or other managerial aspects, there is a converging theme, which deals with how to manage practical tensions, mixed messages and opposition, and how to balance paradoxes within the organization. On a general level this examplified by Sternberg's (2003) confluence model of creativity. He writes:

> A confluence model of creativity suggests that creative people show a variety of characteristics. These characteristics represent not innate abilities but rather decisions...in other words, people decide to be creative. They exhibit a creative attitude towards life. Attributes associated with creativity include (but are not limited to) proclivities to (1) redefine problems; (2) recognize how knowledge can both help and hinder creative thinking...(3) take sensible risks; (4) surmount obstacles that are placed in one's way; (5) believe in one's ability to accomplish the task at hand...(6) tolerate ambiguity, (7) find extrinsic rewards for the things that are not intrinsically motivated to do; (8) continue to grow intellectually rather than to stagnate. (Sternberg, 2003: 391–2)

A recent study by Amabile and colleagues (2004) pinpoints the following behaviours that deserve particular emphasis in the leader's repertoire: (1) skill in communication and other aspects of interpersonal interaction; (2) ability to obtain useful ongoing information about the progress of projects; (3) an openness to and appreciation of subordinates' ideas and empathy for subordinates' feelings (including their need for recognition); and (4) facility for using interpersonal networks to both give and receive information relevant to the project. The study also conclude that leaders who wish to support high-level performance (i.e. support creative action) must pay careful attention to the details of their own everyday, and seemingly mundane behaviour toward subordinates which is the power of ordinary practices.

Sternberg (2003) advances three decisions for creative leaders: (1) 'With regard to creativity, the first decision is that one is willing to defy the crowd'; (2) 'the second decision is willingness to persevere in the face of obstacles'; and (3) 'the third decision is willingness to take sensible risks'. Yet another perspective is offered by Mumford *etal.*'s (2002) tripartite model. This model constitutes a corresponding three-part model for leadership for creativity, encapsulated in the following: Push forward an integrative style – a style that permits the leader to orchestrate expertise, people, and relationships in such a way as to bring new ideas into being. The model consists of three elements: *idea generation*, *idea structuring* and *idea promotion*. One attractive feature of this tripartite model is that it explicitly acknowledges the complex, perhaps somewhat contradictory, nature of creative leadership. A last example is offered by Andriopoulos (2003), who argues that in order to understand the practical tensions, mixed messages or oppositions that characterize the management of organizational creativity lies in how to handle paradoxes resulting from these tensions. Some of these paradoxes could be is the need to learn from the past but seek new areas of knowledge (i.e. 'thinking outside the box'), take incremental risks but break new ground, and encourage diversity but build cohesive work teams. The difficulty often lies in understanding and embracing opposing management practices that can promote organizational creativity, while ensuring financial soundness.

It is unclear whether traditional and available models of leadership can arbitrarily be applied to scientists and people working on complex, novel problems, which would consider what could be called scientific leadership (Hurley, 2003). In fact, Csikszentmihalyi (1999) and Feldman (1999) have argued that success of scientific endeavours is to a large extent dependent on people's understanding of the issues of confronting the field (i.e. other bodies of scientific knowledge in the organization). One implication is that leaders of scientific organizations (e.g. life sciences and biotech) cannot simply retreat into secure isolation of administration, but must instead play an active part in acquiring information about, for example, technologies and trends (Hurley, 2003).

As has been pointed out in the literature, leadership in creative organizations is essentially treated as dealing with paradoxes. Some would argue that the very term creativity leadership is an oxymoron because creativity is, by definition, that which resists any attempt at being subject to leadership. In the following, we will regard creativity leadership as a mechanism for what the German sociologist Niklas Luhmann (1995) calls *de-paradoxification* – that is, the ability to make what is seemingly in opposition or heterogeneous become manageable and intelligible. The notion of paradox is of Greek etymological origin and brings together the two morphemes *para* (outside of, additional to) and *doxa* (opinion) and means what is 'unbelievable' in Greek. The notion of *doxa* has been used by Pierre Bourdieu and Roland Barthes to denote public opinion that is not subject to reflection and public discussion. *Doxa* is here what is deeply embedded in common sense thinking and therefore cannot be surfaced and investigated *qua* opinion. Barthes (1977: 47) writes: 'The *Doxa* (a word which will often recur) is Public Opinion, the mind of the majority, petit bourgeois Consensus, the Voice of Nature, the Violence of Prejudice. We can call (using Leibniz's word) a *doxology* any way of speaking adapted to appearance, to opinion, to practice.' In Bourdieu's sociology (see, for example, Bourdieu and Haake, 1995: 52), *doxa* are the unarticulated shared beliefs that serve to hold society together and help everyday life run as smooth as possible. A *paradox* is, similar to Lyotard's (1979) notion of *paralogy*, what is developed or coexisting adjacent to the *doxa*; alternative ways of thinking and perceiving the world. In European languages, paradox has achieved a somewhat negative connotation as what is absurd or unintelligible and is generally treated as something flawed. This view is criticized by for instance Deleuze (1990b: 74): '[P]aradoxes...inhere in language, and the whole problem is to know whether language would be able to function without bringing about the insistence of such entities. Nor could we say that paradoxes give a false image of thought, improbable and uselessly complicated.' Paradoxes are here not what should be overcome, but rather what are always already being formed on the level of language. Therefore, paradoxes should be recognized as interesting problem formulations that can serve as drivers for social practices and new thinking. Recognizing paradoxes in leadership practice in creative organizations may enable new ways of thinking about leadership and creativity. Again quoting Luhmann (1995), leadership is the temporary suspension of paradoxes in order to enable action. De-paradoxification does not mean to solve the perceived paradox, but rather to make it manageable (Teubner, 2001: 32). In the vocabulary of Andriopoulos (2003), leadership in creative organizations could help the co-workers to rephrase their passions within the general framework of financial performance (paradox 1) or building confidence thorough challenging the individual co-worker (paradox 2). Even though this is necessarily easier said than done, the notion of de-paradoxification is helpful in highlighting the inherent complexity of leadership in creative organizations.

Leadership in creative organizations

A central notion of leadership is the influence that it has on organizational performance. However, there is an ongoing debate about how to lead creativity in organizations. Previous research indicates that creativity and creative processes – and its required leadership – are much more complex than a simple linear relationship between cause (nature of leadership) and effect (useful creativity). Despite the fuzzy nature of creative processes it is still possible to identify recurrent patterns in how the design of organizational structure, incentive systems, cross-disciplinary teamwork and values interact with leadership to support or hinder the creation of new products. These patterns might explain why certain companies historically have been able to achieve repeated breakthroughs. Since middle and project managers have a decisive impact on these aspects, they play an important role in shaping the prerequisites for the flow of organizational creativity. Thus, in order to understand creative actions in organizations, it is important to appreciate leader and team interactions. Therefore, a study was conducted to focus on the interplay between *the leader, the team and the organizational contexts* in order to reveal new perspectives of creative and collective competence in organizations. The objective of the study was to explore different aspects of the interplay between co-workers and management behaviour and styles in different organizational and company contexts in order to improve the understanding of organizational creativity. The cross-company design of the study implies that the knowledge produced has the potential to be useful for all participating companies. The vehicle for this ambition is the actual interface between the companies and their different contexts of leadership: architecture, basic research, and large pharmaceutical R&D. By contrasting these organizational contexts, valuable insights may improve the reflections of management on how to lead creativity in a constructive way.

Companies and respondents

Four companies which have all demonstrated high levels of organizational creativity participated in a research study. The selection of these four creative organizations was based on the notion that all of their primary sources of income come from the generation of ideas, they have gained industry awards, are profitable, or have been able to produce new candidate drugs,[32] and made significant scientific contributions and breakthroughs in medicine, or in the case of the architectural firm, gained international reputation in architectural design.

The questions posed broadly addressed what aspects in leader/team interactions are important to increase the understanding and promoting organizational creativity? But also in what way do leader/team interactions influence organizational creativity. The research was guided by the ambition to stretch the traditional notion of leadership and organizational creativity, and if new insights and valuable knowledge of organizational creativity be created in the interface between different leadership's practices and organizational contexts, i.e. crossbreeding will lead to interesting forms of creativity leadership. The research was conducted by means of semi-structured interviews on each company in order to explore a wide spectrum of aspects relating to one or several projects in each firm. Questions addressed different categories, such as organizational structure and design, daily work aspects, decision-making process, problem solving, risk taking, trust, handling ambiguity and complexity, communication aspects in relation to leadership and organizational creativity. The research involved 30 in-depth interviews with researchers, architects, project managers, senior managers and company directors during 2002–2003. In addition to AstraZeneca (see Chapter 1.) three other companies were represented.

- *ACADIA Pharmaceuticals*, a biopharmaceutical company focused on the discovery, development and commercialization of small molecule drugs for the treatment of central nervous system disorders. The

---

[32] A new chemical entity is a compound that is not previously described in the literature.

company focus on molecule drug candidates using a chemical-genomics platform. Using its proprietary drug discovery platform, ACADIA has discovered all of the drug candidates in its product pipeline. Acadia's headquarters and biology research facilities are located in San Diego, California and its chemistry research facilities are located near Copenhagen, Denmark. ACADIA has more than 100 employees.

- *Carlsson Research* was formed by the Nobel Laureate Arvid Carlsson and his co-workers in 1998. The company is devoted to the discovery and clinical development of new pharmacological treatment principles for psychiatric and neurological disorders. Carlsson Research identifies new drug candidates using an Integrative Screening Process (ISP). This unique drug discovery platform synthesizes knowledge in the fields of chemistry, computational biology, pharmacology and clinical medicine. The process secures beneficial effects at the level of the whole biological system. Carlsson Research has 35 employees working at laboratories in Gothenburg, Sweden.

- *Wingårdh Arkitektkontor* is an architect consultant firm founded by Gert Wingårdh in 1977. The company became internationally known following the exhibition of this building at the 1996 Venice Biennale, and achievements in designing the Swedish Embassy in Berlin, where he expressed original creativity, architectural ability, and a sense of humour which is quite rare among Swedish architects (Wærn, 2001). Recently, the company has won several international invited competitions prizes including the chancellery of the Royal Swedish Embassy in Washington, Ericsson HQ in London, and the Sweden's national science centre Universeum in Gothenburg in 2001. Wingårdh Arkitektkontor has about 100 employees at offices in Gothenburg and Stockholm, Sweden.

The empirical findings from this study have been divided into four themes: *Role of the leadership in organizational creativity, Role of creativity, Motivation and courage in creativity work* and *Communication and practices*, each of which, in different ways, illustrate how leaders and employees in the different firms perceive notions and aspect of management and leadership in relation to organizational creativity.

A note on methodology: narratives of leadership and creativity

In this chapter, we draw on a body of literature recognizing narratives as an important resource in organizations (Czarniawska, 1998, 2004; Boje, 2001). Rather than simply conceiving of remarks from interviews as being self-enclosed statements detached from broader social conditions and organizational and managerial objectives, we here consider the interview to be a situation in which the interlocutor is giving expression to his or her beliefs, received wisdom, or enacted worldviews. In the interview situation, the interlocutor is then constituting oneself as a credible and moral person representative for a company, profession, or social class. In Atkinson and Coffrey's (2003: 116) formulation: '[I]nterviews are occasions in which... "informants" construct themselves and others as particular forms of moral agents.' Therefore, the interview provides empirical material in which the line of demarcation between facts and 'fiction' is not always clear; Gubrium and Holstein (2003: 3) write: 'Interview roles are less clear than they once were... Standardized representation has given way to representational invention, where the dividing line between fact and fiction is blurred to encourage richer understanding.' However, one should not be too suspicious about the interlocutors' objectives, the point is instead that one needs to be aware of the epistemological ambiguities inherent to the interview situation. The second problem is then how words and actions are related. Atkinson and Coffrey (2003: 117) warn that '[w]e cannot take the interview as a proxy for action'. Saying is one thing, then, doing is another. In order to make the empirical material become meaningful to the reader, we are here giving substantial space to the interlocutors' narratives. Rather than citing disjointed passages, the interlocutors' stories and arguments are here accounted for in greater detail than is usually possible.

Role of the leadership in organizational creativity

One senior manager from AstraZeneca pointed out the importance of leadership and creativity, especially in the sense of creating and maintaining a creative climate:

> There is no doubt that pharmaceutical research is about teamwork. In all teams there are occasionally those leaders who can inspire and support intrinsic motivation to others. There are several examples of this. I remember one example from one of our R&D sites in Sweden, in which a number of individuals' unique creativity had a large impact for the development of several of the company's best selling products. However, in this case the most important accomplishment was not their individual creativity, but their capability to inspire and create a sense of team spirit to achieve peak performance. You know there are very few creative persons that can do a drug product themselves; it

doesn't work like that. What is important is the ability to inspire and have the team with you. So creativity for AstraZeneca as a whole is that the team should be creative. There are unique individuals that can initiate this kind of process. In order for this to happen, I believe that one has to let loose things – this creative spirit, and self-confidence are within some individuals. This is also a part of the creative environment. However, in a creative environment, the individuals have to take sufficient responsibilities, and therefore also be motivated to do their best. (AstraZeneca, Senior Management, Discovery)

The role of leadership is given even more emphasis by the CEO from one of the smaller pharmaceutical companies, who made it quite clear that leadership and creativity matters for the business:

Yes, it [leadership] makes a big difference. I don't think that you can have a creative organization if you don't have leaders who are willing to do that. Who are willing to get in there, are willing to continually warrant, continually challenge themselves, because the process will be haunted. And I have seen a lot of these aspects of creativity. I've seen people who have stopped being engaged in the science, who become managers of science and who very soon get pushed far enough away from the science where they can't evaluate what is creative, what is good, what is novel. And so then they start to put barriers on the science, so that the only things that they can handle go forward. If it's beyond their scope then they suppress it. And I think the pharmaceutics industries are filled with that, with those kinds of individuals. You've probably seen them. In new careers as well. You can hit them in the head with an idea for a multi billion-dollar drug and they wouldn't even be able to recognize it. And you see that a lot. You see these structural environments. And I think that structural environments are not there for the scientists, they are there for the managers, but they can't handle the creative process. They can't handle unique and novel ideas and what to do with them and so they set up the environment so that they don't have to do that. (ACADIA, R&D director)

Another aspect of how leadership is being put in practice is taken from a respondent using the metaphor of ice hockey:

First and foremost, the leader is never on the ice. What counts is how the leader puts together the best combination of players depending on the task and opponent. But it is also about preparing the team; get them to know the task they have before them, knowing each other on ice and so forth. You know it takes years to build a hockey team, and you can demolish it in an hour. It is similar with leadership in new drug development. You must also respect the distant players; that not always are seen. As a leader, but also between colleagues, it's about respecting each other's efforts as a part of the game. This sends a clear message that you appreciate the contribution. That feeds creativity. It's all about daring. Speak up about your thoughts. Don't keep them to yourself because maybe you think 'I am not so important'. You know it's much about group dynamics. To create a sense of acceptance, having all ideas on the table – it's like a jigsaw puzzle; some pieces don't fit in, some are not even a part of the puzzle. However, it's more comfortable to take single pieces, but they are not linked together. (AstraZeneca, Senior Management, Development)

Yet another respondent from AstraZeneca reflects on the issue of whether different company cultures may interfere with leadership style in supporting organizational creativity:

Broadly speaking, the Swedish model is very much based on trust and belief that the individual capability, informal, doesn't write unnecessary reports and spending time on administrative work. It was a famous senior researcher in Discovery research in former Astra who once said – as a researcher one should have four days in the lab and on the fifth day you evaluate the result, and during the weekend you plan for next week's experiment. This kind of mentality has more or less been influenced the Swedish R&D sites, especially the site in Mölndal or former Hässle. That is of course a big advantage. When it comes to organizational creativity, you also need a certain structure to be able to catch innovations at the right moment. You cannot only perform experiments, you also need, from time to time, to be formal and reflect what we actually need to do, and so forth. So, I believe that the so-called 'Swedish freedom', in this context, needs to be combined with the formal part of the Anglo-Saxon culture, which has a more structured way of working. If you combine these two cultures one could add value. Maybe the Swedish model works better in Discovery compared to Development, because the English culture is more regulatory dependent. (AstraZeneca, Senior Management, Discovery)

On the issue of whether leadership and its implications for organizational creativity is formalized or pronounced, one of the senior managers from ACADIA responded:

So fundamentally I guess I don't understand the process well enough to manage it and I would be afraid that anything that I do would screw it up. So I try not to manage that process. When I have somebody who is a really creative scientist, I try to be supportive to make sure they are given the time to do things that they need to keep them excited and moving forward. But it's beyond my management skills to say: How can I take that other person who isn't doing that and make him like this other person? Is it because it's fundamental to the individual? Perhaps. There are creative people and there are other people who are much more likely to just follow down the steady path. There are people who like to think broadly and innovatively. There are other people who would say: 'That scares me, and I want you to tell me what I need to do and I'll do it, but don't ask me to find my world myself'. I kind of think about it, you know – when things intrigue me – when I was watching the Soviet Union disintegrate, was that a person said: 'Everybody is going to be excited. Free at last. And I can do whatever I want to do'. And that wasn't the case. There was a good portion of people who said 'Oh, my God, I'm free; I don't know what to do. No one is telling me what to do. I can't take this any more. Somebody come in – let's bring the communists back because they are going to tell me what to do. And I am very comfortable if somebody will tell me what to do'. I think the same things are present in science work. There are people who just love being told what to do and want to follow a set pattern and be comfortable with that, and there are other people who you try to do that to, you would kill them. I think that one of the goals in management should be to identify those people, separate them and treat them differently. We do recognize the people who are going to be innovative, who are innovative. The fact that people who are innovative, being creative are the ones that are put into positions where that's a useful characteristics. People who do not have natural capacities to be creative are put into other position where they are still valuable to your organization... and I try to do it from a role model standard. And try to convince people that you can be creative and still be in an industrial setting. (ACADIA, R&D Director)

On the issue of how leadership style and creative climate influence organizational creativity, one senior manager from AstraZeneca explained:

What is optimal leadership behaviour in order to stimulate a creative environment? Basically, I believe it deals with how confident you are as an individual and in the role as leader. How much do you trust you colleagues and team? What needs you have for getting into details? Your capability to inspire? In essence: Being secure in the role as a leader that is the most important thing. (AstraZeneca, Senior Management, Development)

One of the dilemmas of managing organizational creativity and long termism is exemplified by a senior manager in ACADIA:

First, you have to be open for new ideas. The new drug development process is often given the form of a diagram were you have a long series of interconnected activities that start on a certain point and then ultimately ends with launching a product on the market. The interpretation of the modern drug discovery process is that you begin with target identification, target validation, screening, chemistry identification, lead generation, lead optimization, and so forth. This way of performing research came from the notion of the rationalization of the research process. So, this means that you begin with spending a tremendous time on evaluating the receptor of interest. This means that many of the large pharmaceutical companies spend a long time on evaluating their targets. It can take up to two years, and that part – the two years – has been added to that time you previously had on a drug development project. So, instead of using new technologies and integrating them in the existing process, you put them in the beginning of the process, instead of shorten the process. This is, in my opinion, quite stupid. I'll give you an example. Half year ago, we had an employee who came from a large pharmaceutical company working in a target validation group. He was tremendously proud of what they have achieved and presented what they have been working on during this two years period within a group of seven highly qualified researchers. After two years a decision was made that resources for performing the subsequent pivotal experiments of the target were not available. So, to be cynical, during those two years: Seven persons, large research capacity, and budget to think about of what working at all on certain target. That is what I call wasted resources. (ACADIA, CEO)

Yet another example of the importance of leadership in achieving a balance between the combinations of tightly and loosely coupled systems ensuring that they are never linear since such processes are complex and complicated to manage and therefore need continuous understanding and awareness. As a consequence, organizational creativity may be regarded as an impermanent or volatile factor to consider. This example is taken from a researcher in one of the smaller pharmaceutical companies:

> Among these semi crazy academics it got to be some kind of control. Otherwise too much time is spoiled in academic quarrels and debates which cut us off from the question of fact, and becomes more centred on matters of prestige which is counterproductive. So it is important with management control and leadership. However, it is a big challenge because it needs understanding. It wouldn't work just taking someone walking in and leading this research team as any ordinary team. It is like riding a horse, you can't be to firm, that's the real challenge, because then the horse won't move. And you can't give free rein because the horse will run into the woods. It demands a kind of instinctive feeling combined with understanding of the research process. (Carlsson Research, Researcher)

The balancing between stabilizing (i.e., management control) and destabilizing (e.g. to what degree do researchers allows to solve problems and take actions that are most suitable in a given situation) mechanisms in leadership work is pointed at by one manager at Carlsson Research:

> It's a kind of balance between participating, and knowing without controlling too much, but yet influencing and dealing with external company issues. (Carlsson Research, Manager)

Wingårdh Architecture Firm provides an illustration of how leaders become involved in creative processes in practice. In architecture work, the process including idea generation, preparation and different design phases is referred to as the *sketch process* or *sketching*. The sketch process is regarded as one of the single most creative events in the architecture's work. The background for this example is the preparation and design of the national science centre, Universeum, opening in Gothenburg in 2001. The building is more than 10,000 square metres in area and contains aquariums, experiment stations, exhibitions and IT-based education solutions. It is notable that wood, glass and concrete are the three main materials employed. If at some time in the future the building has to be knocked down, this is made easy by the construction solution and all the building materials can be recycled. Universeum has been built to be an ecological role model in every way. The architect responsible for the project explains:

> It can be a 'sticky' process – that house should be there, another there; the communication should be so and so. In the case of Universeum, the first idea was a collage of different parts and a connection with different landscapes – the rainforest was to be connected with the actual exteriors through a large set of glass window. The key point was that the perception of the rainforest should be larger than it actually was when contrasted though the external surrounding of the building. This idea came early because the surroundings were good. Then different parts emerged but details hadn't yet found their right shape... I presented the second concept to Gert [Wingårdh] and he thought it may work. He then worked it over and faxed it back to me combined with the new concept of using large pieces of wood in the building. This process included a number of iterations back and forth. The idea of the pronounced large wood entrance derived partly from the environmental aspect, which was already described in the competition requirements. Our first intention was to build the large roof by using an ordinary steel construction. But in the continuing sketch process, the concept of using wood not only for the entrance but also for other parts emerged. This was heavily influenced from an idea Gert got from an old castle he had visited recently. Then we decided that a new point of departure would be a building with very general and simple design with large flexibility. And this fitted into the wood concept. We wanted to show that it was possible to build a modern building in wood, instead of concrete, something that is actually quite rare nowadays. This wood construction was actually much more expensive than steel, but it represented more originality, and was in line with the ecological objectives. Although more expensive at the time of construction, in life cycle perspective it will become cheaper. (Wingårdh Architect Firm, Architect)

A senior manager in AstraZeneca emphasized the importance of mastering a field of expertise when acting as a leader:

> You need to be able to talk the same technical language. It is a quite unreasonable task for a leader in new drug development to match the best persons in different scientific domains. It doesn't work like that, but you need at least to understand the broad picture of what they do. But more importantly to be able to understand, at a high level, what prerequisites are needed to perform of good research. Because that process – to conduct qualitative research – is shared between different scientific domains. That is to say, being able to evaluate the research output, know how to work up that material, know how to present it; those are the leader's primary skills. The other issue is that you need an organization that can secure and support required leader competence and enable new thinking through substituting competencies. Specialists need to be able work across disciplines. Leaders need to be respected and demonstrate an ability to represent his or her own function in different projects. (AstraZeneca, Senior Management, Development)

On the rather specific issue of how much time the leader should spend on the dialogue, and on being an interactive part in the work with subordinates in order to support organizational creativity, the study displayed a rather wide range of responses. The averages were as follows (including leaders and employees): ACADIA, 30–60 per cent, Carlson Research, 30–40 per cent, Wingårdh, 50 per cent, and AstraZeneca 20 per cent.

Role of creativity

On the rather broad issue of what role creativity plays in the different companies, all respondents claimed in different ways that creativity has a major – albeit complicated – influence for their business. The CEO outlined the role of organizational creativity in the case of Wingårdh architectural firm: the CEO explained:

> I believe creativity is to come up with a solution that in a surprising way solves the problem. We have an internal motto – our mission is to give the client what he didn't know he wanted to have. This is to indicate that something beyond the standard solution is needed for a certain task. To be firm, I would say that architecture is an innovation profession not a creative profession, because we are constantly

manipulating known constituents and elements. Being creative; would be to take away the pillars making the house float in the air – we don't do that very often, especially not with good results. Architecture is a craft with something extra. Architecture, as domain, quite seldom develops new materials. Architects in general are not very aware of different construction materials, or deep domain knowledge in technology aspects. (Wingårdh Architect Firm, CEO)

The companies studied here recognize the importance of allowing employees to capitalize on their creative endeavours. Employees are encouraged to identify the norms or gaps associated with a problem or an opportunity identified and then breaks them down and tries to highlight any areas for potential innovation. A senior manager in AstraZeneca points out that creativity is to a large extent dependent on a collaboration with different parts of the organization:

For me creativity is to think different than others. If one thinks as everybody else then you are not creative. If you comparing drug research today with 25–30 years back, we are heavily dependent on multitude contributions from different individuals in order to succeed all the way to a final product. It is group creativity or organizational creativity that is more important rather than individual creativity. It is the wholeness that is important. (AstraZeneca, Senior Management, Development)

Another aspect of the crucial role that organizational creativity plays in ACADIA is explained by the CEO:

Without organizational creativity we will simply die. I guess that's counts for all small companies. Maybe that is more pronounced in the US where the business climate is tougher compared to Scandinavia. Creativity for us is also needed in search for businesses' solutions, finding or partnerships and collaboration with other firms. It takes a lot of creativity to achieve that. Then our way to combine external with internal focus demands for creative solutions. We simply cannot afford to have slack in our organization. In order to succeed for a company like ACADIA, we constantly need to optimize our resources in the best effective way. (ACADIA, CEO)

For a researcher in Carlsson Research the role of creativity is about understanding the underlying principles in the research:

Creativity is about to find and understand a new effect, principle and the underlying mechanisms. How do things relate to each other? For the chemists, it is a leap – a new way to synthesize molecules – what should they look like? But foremost, being able to recognize the new, and understand. If you have been working for a long time with a bunch of molecules that you don't understand completely, and then be able to take that leap: No but stop here for a sec: – this is a new profile, a new concept. (Carlsson Research, Researcher)

Another aspect of organizational creativity in the case of the Wingårdh Architectural firm is given in the following:

I would say it [creativity] is highly important. It is creativity that drives [the business]. In a sense that's why we are working and our CEO is very firm about this and communicate this clearly. Why should one design buildings if one is not creative? I mean, one cannot design the same kind of building over and over again – that would just be replication. That would not be fun or challenging. (Wingårdh Architect Firm, Architect)

For a researcher in AstraZeneca, the role of creativity is the ability to use existing knowledge in order to create new chemical compounds:

Creativity in the context of the pharmaceutical industry means the ability to put together the available knowledge to understand or to propose things that other people have not done before. It must always be based on accurate knowledge of the specific domain of science or therapy. (AstraZeneca, Researcher)

This more scientific view – knowledge had to be 'accurate' and 'specific' – comes from demands on researchers to deliver new candidate drugs. Creativity thus becomes the capability and the continuation of mastering a particular field of expertise. Yet another researcher in AstraZeneca argued that creativity is the capacity for holistic thinking, which enables an overview of an area:

Creativity is the ability to see the big picture; being innovative is to identify and find solutions for important problems. For example, if you have a flood in the house, an innovative approach would be to change to wider pipes. A creative approach would be to reconstruct the layout of the house. (AstraZeneca, Senior Manager, Discovery)

Another respondent said that creativity is finding and developing a drug that can satisfy a medical need, which involves interfering with different agendas:

> Creativity is to produce a product that is better than what we have today. We need to think more broadly. I can understand that some scientists don't always agree on this – he or she maybe has the goal of finding the right gene sequence or finding new ways of synthesizing a certain molecule – which may be very creative as such. (AstraZeneca, Manager, Development)

Here, creativity is the ability to align identified problems and market opportunities with existing knowledge. Creativity is thus context-based thinking that broadens the scope through radical rethinking of what is at hand. Finally, one of the researchers in AstraZeneca referred to creativity as a cultural quality:

> It's quite easy to say we want to be creative, but it's very difficult to generate a creative culture. It is not an easy thing, I mean: anybody can say that, so they used the word innovation rather than creativity then because innovation can be managed. Can creativity be managed? Only high quality managers who understand creativity can manage it. (AstraZeneca, Researcher, Discovery)

Motivation and courage in creativity work

Quotes from managers and co-workers display an array of perspectives on the extent to which – and the way in which – different leadership traits and motivation play an important part in order to support organizational creativity. This example of the importance of courage is from a respondent at AstraZeneca:

> New drug development is much dependent on how to separate our products from existing products. That is creativity. To have an on-going discussion on for example if we document this product this way; is that what we always has done or can we do it another way that can actually result in a new product that can distance our competitors. To be creative is also about having courage. To be

able to think 'outside the box' of conventional methods of research, test new limits, and the industry not always very successful in this. However, I believe a cornerstone should be to have communicate – what kind of culture do we want in research and development, and that is to have a creative culture and being able to nurture and support that culture. (AstraZeneca, Senior Management, Development)

Another respondent gave this example, which points out that organizational size can also be an obstacle to creativity:

I believe it has importance. First and foremost it must be a leader, manager that has the courage to release creativity in the organization allowing employees being creative. I think it is a question of self-confidence of the leader. It also helps to be a small organization like ACADIA, which means that it is easier to communicate, with less bureaucracy, to involve the entire organization in the process. (ACADIA, CEO)

Another manager at AstraZeneca describes the organizational creativity as interaction between the leader and co-workers and and its strong emphasis on having a dialogue and openness:

It is interplay. You need to have individual creativity in order to have organizational creativity. However, individual creativity doesn't necessarily lead to organizational creativity and that depends a lot on what kind of values resides in the organization. Working in teams, in our case drug development projects simply demands interplay between different competences. This interplay, if it only concerns that everyone contributes on his or her part, is then a process with a constant speed. But to build in the next dimension, which are quality, creativity and innovation to work – then dimensions like values, success, team, information and knowledge sharing become necessary. It is in this aspect the leadership and the relation leader/co-worker is decisive. I would like to see leadership as a dialogue with co-workers. You cannot see leadership as an isolated trait within a number of important individuals. Instead it is the dialogue that is important and the kind of values management communicates, basically believing in openness, stimulating and rewarding teamwork. You have to permit time for exploring work, because management believe that this in turn will lead to a more long-term effect on the business and that will pay of in the long run. But you should also be careful of using short-term metrics of efficiency by minutes to actually be able to see that there is a bigger picture. Equally important is leadership that communicates clear goals and priorities, and can talk openly about the need of success and productivity, but in a way that can act as a navigator for organizational creativity. It is when you become to detailed in goals, methodology, strategies and so forth; the creative process is negatively influenced. This is about a dialogue, and the language needed is unclear and rambling. (AstraZeneca, Senior Management, Discovery)

Several studies suggest that motivation is an important component in organizational creativity. This notion derived from the *intrinsic motivation principle* of creativity, which suggests that people will be at their most creative when they are primarily intrinsically motivated, by the interest, enjoyment, satisfaction and challenge of the work itself. Intrinsic motivation is relevant to the interest and curiosity in science and high motivation. The component includes factors such as motivation, joy and curiosity attitudes that influenced the research or characterized the project work or correlated as success factors for different projects. One respondent examplifies the leadership aspect of being able to promote this aspect:

In order for you to provide intrinsic motivation for creative scientists, you got to be able to communicate what they are working on. And if you have 15 creative scientists, they usually are doing 15 creative things, so you have to be conversant in what they are working on so that you can engage them scientifically. But that's part of the intrinsic motivation. You say – that's great, go back and come back to me when you've got a real compound or something like that. That is not managing a creative process. And I don't think about managing a creative process. But as you're forcing me to think about it then I think what it is from a certain point of view, you got to be engaged scientifically at a fairly deep level with all the various aspects so it's a challenging proposition. I can understand why lots of managers don't go down their platform. Because it is very easy to manage you know, by objectives, by goals and so on. And that's why lots of organizations build those in. But I think that takes a creative process. So it's very simple to see why it's not very simple to understand why those persons for the calcium receptor, weather or not that has any therapeutic benefit and how can you – to steer that process down throughout the pathway for scientific discovery to commercial. (ACADIA, Senior Management)

Yet another example the strong link between intrinsic motivation and organizational creativity in pharmaceutical research:

> There is something I think is important but somewhat misunderstood, and that is what really motivates people, and that's to work on something that have a medical need or a specific problem – There is not an existing drug product available, or the existing ones have side effects or something else. There is an altruistic side in all people who stimulate in their work on drug research, therefore one should try to get people stay in the project during as long as possible, follow the different phases. Because I think that is stimulates creativity, and then be able more strongly use their increased knowledge. Initially, we are all unfamiliar with various topics that we work on, and then one has to learn. You get automatically a commitment, and commitment feeds creativity, then you begin to think on issues and problems outside work and so forth, and ideas begin to take shape because it is fun. That's why I think the company should do much more to preserve continuity, and not moving around people in projects too much, and not having project leaders who doesn't understand. (AstraZeneca, Senior Researcher, Discovery)

The following example, taken from one of the smaller pharmaceutical firms, reflects the leadership aspect in two ways – the ability to produce good illustrations, images of creative action in the organization, but also to accept failures:

> Well, I think there is some very new thing that hopefully the leadership does make any difference. We screen very, very broad and there are a lot of things that get exciting in interesting chemistries but don't see the light of day, because the management hasn't identified those as exciting kinds of things. So, nuclear receptors, another area that this company never started, nor thought that terminology would be easy. So one scientist tried them, figuring out how to make it work, and has discovered the first ligand for a receptor, a receptor involved in diabetes. Other companies have been working on it for ten years. And the conclusion is there is no way to activate this receptor. It can't be done in fact. This person was on the podium of a symposium and a guy from Glaxo got up

before him giving a talk on of – showing how it's impossible to find a ligand for this receptor. And to the work of the scientist who might already know that his presentation was going to be on the characterization of this new ligand for this receptor. So there is the situation. If you sat down with the literature you never would have tried this. If you would have taken the classical approach which was to do a screening experiment and so on you never would have found it, because the interactions are complex and multifactorial. The only way you would ever find this ligand is if you would have taken the approach this scientist did. And then once that scientist had that ligand[33] it was a situation of the management, scientific management said – My God, this is the Holy Grail in diabetes. Let's keep going on this. Let's drop what we are doing on these other things and focus on this cause this is the most exciting thing from our perspective. And that scientist has a lot of visibility in the organization. Two years ago I think you could ask ten people in the organization – Now everybody – He's the guy that's spoken on the nuclear receptors. And that area has continued to expand, because now it seems to benefit from what happens when you make these kinds of successes. It fed successes in other areas as well. So I think that's a kind of creative example. Hopefully that will be to some productive promotion led to the first ligand for a nuclear receptor. Well, I think what the leadership did here was to facilitate it, to say – yes, your creativity has value, and this is why we think it's valuable. Keep doing it. Do more of that. Even though it was completely out of scope. Completely out of scope. Yes. In fact we went out and got consultants, we thought reading in the literature that this was exciting. We would not have gotten there and brought them in and had brief presented data to them and the said – This is exciting! So that even reinforced it more. We didn't? Our internal expertise, but we got the world authorities on diabetes to come here and see what we are doing to see if we thought we were on the right track. And they were excited about it, so of course that gave the chance to present it to some of the world's expert on diabetes, but I'm sure it was excitement for him. (ACADIA, R&D Director)

---

[33] Ligand: a molecule, ion or atom that is attached to the central atom of a coordination compound, or other complex. For example, the hydrogen atoms in ammonia (NH_3), or the ammonia molecules in $[Co(NH_3)_6]^{3+}$ are ligands.

Another illuminating example of being able to inspire through leading by example is given below:

> He [Arvid Carlsson] has never been poking in details, he allows quite a large freedom among researchers, occasionally he steers up things if he believes things is going out of hand. On the other hand, he comes up with ideas, and sometimes running quite radical experiments all by himself, which on several times resulted in entire new concepts and methods to work with. What I mean is that multivariate analysis combined with in-vivo screening, which is the way we work is quite unique in the business – also reflects our philosophy of doing research. (Carlsson Research, Manager)

Yet another example from one respondent displays the importance of personality and ability of the leader to inspire and to commit co-workers to sharing common visions:

> He is the sort of person who could carry you into a swamp because of his personality. He could persuade you that this is worth going across to get to the golden land of the other side. You may not be able to see the land, but I think you'd probably step into the swamp. I do believe that leadership and all of these things are true, but it's about people having confidence in you and believing in your sincerity and believing that you're somebody worth following. And it's worth taking a risk in committing themselves to you. (AstraZeneca, Senior Manager, Discovery)

In the next extract a senior manager from AstraZeneca gives an example of a leadership paradox dealing with the ability to reflect, and act on how project work can continue to deliver what is expected, while taking new initiatives to avoid conformity – encapsulated in his reference to the 'comfort zone':

> When daily activities become a routine then you are in the comfort zone – and that is dangerous. There is a risk that we become in a treadmill at work, because you do that work that is expected; that doesn't add value. Instead it should be about how individual creativity can be combining with others – the wholeness of creativity that counts. That's for me the most important thing for me. (AstraZeneca, Senior Management, Development)

One challenge for a large, complex organization such as AstraZeneca is to keep up with financial growth, and be cost-effective while still remaining innovative. One respondent reflects on the issue of maintaining a balance between intrinsic and extrinsic motivation:

> You know we all have agendas – an important thing, and not everybody reflects on this, is that there are many reasons to what and why you work, and the best output the company can get is when the personal agenda can coincide with the company agenda. I have an extensive experience of hiring people in this firm. The reason I began to work in this company was because I wanted to develop medicine; that was the goal. In a lot of job interviews during the years, I have asked what's your reason for work in the company, what do you want to achieve? Many answers have been because it's exciting, working internationally, good career opportunities, and lots of other reasons. There have quite few persons who want to work because they want to develop medicine. I believe we should hire more of those people whose personal agenda overlap with the company agenda. (AstraZeneca, Senior Researcher, Discovery)

Communication and practices

Organizational creativity is not simply an instrumental process of aggregating ideas and concepts that may become potential innovations; it may also include new ways for organizations to use images, narratives, and languages for communicating viewpoints. This section considers different factors that will support organizational creativity in the different companies in our survey.

Most respondents emphasize the importance of a dialogue and communication skills in the creative process. One factor was the way in which intuition is increasingly affected when planning and managing drug discovery research. This is illustrated by one of the respondents:

> Creativity is within the dialogue and conversation, it is a communication between the image and building. It is the architects who should design the house and it is me who should design the image, so I am creative when I communicate discuss things with the architects and other in the project. (Wingårdh Architect Firm, 3D-visualizer)

One of the respondents describes this dialogue as a complex iteration in the creative process:

> It's like table tennis. You cannot say that it is a single individual who solves the problem; it's rather a complex collaboration, which leads to solutions. If you have a problem, you divide it into different parts; try to get so much information as possible. I cannot say that I alone have been creative; it's more like a slow process. This is for example reflected in patent applications and publications, where many people are directly and indirectly involved. And from my point of view no single individuals is pointing out; it's teamwork. (Carlsson Research, Manager)

Yet, one respondent gives that emphasizing the leadership role and ability to communicating another example:

> The leader should stimulate a dialogue, a debate; one should not to force things down the throat to people. Being able to get individuals to experience things themselves stimulates to creative thinking. (AstraZeneca, Senior Management, Development)

A senior manager in AstraZeneca pondered whether traditional generic leadership programmes are designed to address the leadership of organizational creativity:

> I think leadership programmes in general, including those we have in AstraZeneca, are too much of the conventional view of leadership, for me it's more about old fashion management. When I participate in these introduction programmes for new managers in all parts of the company, I am careful about not performing a conventional type presentations: this is my organization, this is how we organize things, here are the processes, and so on. Instead, I try to engage people in round table discussions concerning behaviours and talking about what should you do in these situations? How do you look upon yourselves and others? What can you do today, and tomorrow? Basically, I believe we should try to find a good way to change the culture, and that's not easy. This kind of self-conscience, and knowledge about you and the environment and about others, is what really is important. It turns up quite often that this kind of large organizations; there is little knowledge about creativity and

these kind aspects. Instead one uses processes and formal procedures to cooperate that's doesn't feed any creativity. (AstraZeneca, Senior Management, Development)

Innovation and creativity are ambiguous concepts and may thus end in oversimplification and ambiguity, which may become meaningless for the organization; there is a link to the context of the concept and related processes, and the individuals behind it (Sundgren, 2003). Even though organizational creativity is perceived as a highly prized organizational capability in pharmaceutical research, it is something that is hardly discussed in daily work (Sundgren, 2004). An ongoing narrative on organizational creativity would facilitate more effective research practices. In this perspective the case of Wingårdh Architectural firm actually have a vocabulary, and speech genres for communicating organizational creativity in the concept of 'sketching'. One architect explains:

We talk about sketching – one sketch a project. This is the creative process in a way, in working with the project in different phases. We do it a lot – try to select those things are important and combine these in new ways. All demands cannot be put together; these things work, these don't – it is a puzzling. It's like rock climbing, how should I attack the mountain, what is the best possible route, conditions of the mountain wall, what are the possibilities to succeed, what techniques should I consider and so forth. You know, were should the entrance be, parking lots, how to get the right light into the building, how does that affect the canteen, how many floors and so on. What I mean is that there a lot of factors involved, and they are unique for each project. Then the sketch continues throughout the project. You have a concept of how the interior should look like, then that changes, you take ideas from previous projects. This creative process is there all the time. (Wingårdh Architect Firm, Architect)

Much of the research literature on creativity has put great effort into describing what creativity is. More important, but less explored, is the question of *where is creativity located*? Previous research has shown the importance of social networks for promoting creativity, especially the form of informal networks that are similar to 'weak ties' (Granovetter, 1982). Informal networks in this context are different types of communication, social contacts, interactions, and information exchanges that occur outside an employee's ordinary line and project organization, but useful to different research projects. Thus informal networks seem to become important entities in which creativity seems to be hidden in an organizational capacity of *connectivity* (Perry-Smith and Shalley, 2002). This aspect, but also the lack of it, is pronounced by one of the respondents in AstraZeneca:

> I think networking is a very undervalued and underused resource. The business should be doing far more to encourage networking (and this is very difficult in the current budgetary climate). It is my experience that networking creates a really positive can-do attitude, particularly across sites, when people are not working in an "us and them" environment. (AstraZeneca, Manager, Development)

Yet another example of the dilemma between being able to maintain informal networking vis-à-vis workload is exemplified below:

> When work load is increased maintenance of these networks becomes compromised and has a longer term detrimental effect on productivity, e.g. the network no longer exists due to lack of participants time and when you really need to tap into some advice/experience it is no longer there. (AstraZeneca, Manager, Development)

Another respondent stated when asked about how informal networks are used within his project:

> The organizational structure of today might not facilitate this project. We have to be on our guard so that we can see that this model that our work originates from [how work is performed, through which methods] is established. It has been build up pretty much around networks, personal networks. It is not the organization that has given us the tools to start up this project; it is our own initiative – that is how it is today. (AstraZeneca, Manager, Development)

A researcher from Carlsson Research exemplifies the importance of being able to easily receive and share information, which highlight that information is the backbone of a creative environment:

> I can talk with the chemists sitting next door, or pharmacologists, lab assistants, or Arvid [Carlsson], this communication becomes very vital and dynamic because we are sitting so close. It is the short

distances and the high information transfer that makes it so easy to get to know the things that is important. (Carlsson Research, Researcher)

Leading creativity: de-paradoxifying opposing objectives

What do the statements in this chapter tell us about management practice, leadership and creativity? What do they reveal about organizational creativity in new drug development and in architectural work? The examples illuminate some important issues, questions, dilemmas, and ambiguities relevant to understanding and managing organizational creativity. The image or role of creativity in these companies reflects the idea that creativity in organizations involves divergent and convergent thought and action before it becomes effective, but also that organizational creativity is the outcome from collaborative efforts. Creative actions are constantly influenced by complex interaction, overlapping domains of knowledge, and different stakeholders. In the case of new drug development it is not solely about delivering new candidate drugs, but also includes all the activities in the pharmaceutical industry: new drug development activities, strategic management decisions, and human resource management practices. This aspect is even more pronounced in the case of the architectural firm. So organizational creativity encompasses the entire organization and consists of a multiplicity of activities. Statements show a rather broad and multifaceted image of what leadership behaviour is about when to support creativity in their organizations. However, some important leadership aspects emerge. These examples support previous research on important leadership traits to support creativity such as: the ability to inspire, good communication skills which in this case means to be able to perform subtle, yet a dynamic dialogue with members of the organization, and above all to have courage and self-confidence.

The most important aspects of this study deal with management's capacity to handle different paradoxes. But also leaders willingness, and ability to understand what organizational creativity is, and what it means for their organizations, or as Andriopoulos (2003) argues to get leaders a sharper view of the dynamics within creative environments. Previous research in the organizational creativity area, display absence of integrative explanations by the fact that studies have approached the area either conceptually or by testing hypotheses based on others' findings or writings (Andriopoulos, 2003). Furthermore, their research outcomes tend to be descriptive by focusing on 'what' is happening in the issue in question. This study illustrates that the management of organizational creativity is a much more complex process. Further research into the study of paradoxes should therefore be encouraged as they provide a holistic and dynamic view of the phenomenon under investigation.

One fundamental aspect of organizational creativity per se is to challenge the status quo, and it is therefore potentially problematic for an organization. In management practices it will undoubtedly challenge conventional (inherited) rationality, which management considers effective; consequently, organizational creativity becomes problematic. It is therefore important to balance the focus of managerial rationality and become more directed towards acknowledging management practices relevant to understanding and promoting organizational creativity. This leads to a need for management to adopt new thinking that will enable a less rationalistic view and acknowledgment and understanding of core and layer creativity in new drug development.

A model for management practice, leadership and creativity

One way to illustrate what a new thinking in management practice means for the support of organizational creativity is to use the concepts of stabilizers and de-stabilizers. Organizations have many stabilizers but quite often lack proper and functional destabilizers (Hedberg and Jönsson, 1978). Stabilizers are, according to Hedberg and Jönsson (1978), established fixed repertoires of behaviour programmes over time, and many become too rigid and, therefore, insensitive to environmental changes. Stabilizers filter away conflicts, ambiguities, overlaps, and uncertainty; they suppress many relevant change signals; and kill initiatives to act on early warnings. Destabilizers represent organizational factors, or behaviours that are dynamic and unpredictable; they challenge the conventional and thus destabilize the organization. The logic in use for managing projects in, for example, new drug development is substantially influenced by instrumental rationality, projectification, and planning. The goal is to ensure processes that enable projects to deliver high-quality output in a timely fashion. These aspects can be called stabilizers in the sense that they improve efficiency and ensure uniformity, reliability, and predictability.

In the case of new drug development, several drivers for organizational creativity can be defined as destabilizers, such as informal networks, information sharing, new skills (e.g. rational persuasion, political entrepreneurship), and intuition (Sundgren, 2004). From an organizational creativity perspective, proper destabilizers are important to promote creative action. It is important to note that the model in Figure 7.1 does not argue about the need for stabilizers. The basic argument is that organizational creativity in many for example in new drug development is too embedded in stabilizers (projectification, planning, and so forth) and needs to move towards practices that make better use of destabilizers. In this sense, new management practice and strategy are necessary to understand and create a balance between the two systems – to promote organizational creativity. The double arrow in Figure 7.1 suggests that management should not only bridge the two systems and secure a strategy in which productivity is not enhanced at the expense of organizational creativity, but also aim to create a kind of creative equilibrium between two systems.

Traditionally, there is a strong urge for leaders to acknowledge stabilizers. In fact, leadership and management skills are often defined in terms of stabilizers. The primary reason for this is that it is a safe arena.

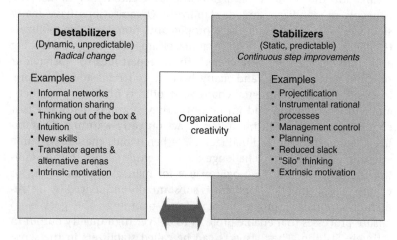

Leadership and management practices

Figure 7.1   The creative equilibrium model of new thinking in management practice to support organizational creativity in new drug development

The leader can apply already accepted rules and behaviours, which include to a large extent the adoption of predictable control mechanisms to evaluate established success factors and rewards. A leadership style that is based on an extreme adoption of stabilizers also reflects insecurity and a flight from ambiguities. The other extreme is a leader that is always in the destabilizing area. This leader continuously challenges the established routines in the organization. This style may have some resemblance with the classical entrepreneur, such as boldness and strong belief in changing the organization. Subsequently, this kind of leadership may come into conflict with the organizations vision and goals or alienate co-workers over time. Put into an extreme; one may argue that the leader does not understand, or lack to comply with the direction of the company, which may too much threaten the organization.

From an organizational creativity perspective, the model suggests a leadership style which can create a dynamic balance between stabilizers and destabilizers. Through an ongoing reflection and understanding, on the direction (vision and goals), at any given time, the risk of imbalance between stabilizing and destabilizing can be minimized. One challenge is thus to be capable to handle both polarities at the same time. This leadership do not only have condition to support and nurture organizational creativity, but also to increase efficiency with enhanced intrinsic motivation. The only effective way to create the right balance between stabilizers and destabilizers is through effective communication and dialogue about the firm's vision and goals. Subsequently, this leadership will open up for new thinking, change and supports revision of control mechanisms. Another consequence of the model is that it demands a leadership style that is less prone to controlling everything, but to mediate or lead though others. Thus it needs a dynamic, yet subtle balance between the two dimensions were leadership and management practices are characterized by The adequate knowledge of the scientific process, insight of organizational creativity, and courage.

Conclusions

By applying a *systems perspective of creativity* it is possible to decrypt the code of the dynamics involved in managing organizational creativity, which can open up for a more effective management and leadership, and give leaders a sharper view of the complexity creative environments. There is no doubt that management and leadership have a considerable influence on creativity in organizations. However, in order to effectively and constructively manage organizational creativity demands a capacity to handle different paradoxes and opposing forces. This demands a capability to recognize, support and handle destabilizers and stabilizers in an organization.

8

Conclusion: Managing Organizational Creativity

Introduction

The final chapter of this book will examine some of the practical consequences of the topics discussed in the first and second parts of the book. In the first part of the book, the literature on organizational creativity was explored in terms of its theoretical consistency, methodology and epistemological considerations. The second part of the book pointed to issues such as the influence of technology, leadership and cognition as areas that the organizational creativity literature could make more explicit reference to. In summary, the first two chapters of the book provided a comprehensive critical review of the organization learning literature *qua* theoretical field and a subdiscipline within organization theory and management studies. In this last chapter, we will take the consequences of these theoretical concerns into account when examining the practices of organizational creativity management. The main source of empirical illustration and observations is the pharmaceutical industry while there are a few additional studies and examples in the literature that are referenced. In this chapter, we draw on Peter Drucker's classic text *The Practice of Management*, first published in 1955. In this book, Drucker emphasizes the 'creative' components of management and defines management in more entrepreneurial terms than do the contributors to the leadership discourse. Drucker writes:

> What then is 'managing a business'? It follows from the analysis of business activity as the creation of a customer through marketing and innovation that managing a business must always be entrepreneurial in character. It cannot be bureaucratic, an administrative or

even a policy-making job. It also follows that managing a business must be creative rather than an adaptive task, the more a management creates economic conditions or changes than rather than passively adopts them, the more it manages the business. (Drucker, 1955: 39)

Drucker then tries to save the notion of management from being overly associated with the bureaucratic organization forms, an organization form that has from the outset been subject to detailed and perennial criticism in terms of failing to adapt to external changes and new conditions (Starbuck, 2003). For Drucker (1955), management is what is creative and explores opportunities. In a more contemporary vocabulary, one may say that Drucker stresses what Teece, Pisano and Shuen (1997) have called *dynamic capabilities*. Others have defined dynamic capabilities accordingly: 'A dynamic capability is a learned and stable pattern of collective activity through which the organization systematically generates and modifies its operating routines in pursuit of improved effectiveness' (Zollo and Winter, 2003: 603, the original in italics). Seen in this view, management is what orchestrates an ongoing adaptation to external changes in order to exploit opportunities. In other words, when we speak of management in the following chapter, we do not only denote the transactional leadership practices, the day-to-day administrative concerns, and the routine-based work, but equally the development and use of the firm's dynamic capabilities within areas that are determined by the organization's creative resources.

The use of narratives and storytelling

One of the most important competencies of the manager in an organization which relies on its creative capabilities is skills in managing joint narratives and the sharing of experiences (Ready, 2002). Narratives are stories that are 'emplotted', that is, when events and occurrences are given a storyline with a beginning, middle and end, and have a meaning and a moral that can be shared by the storyteller and his audience (Czarniawska, 2004). Since at least the publication of David M. Boje's (1991) study of narratives in an office supply firm, the notion of narratives has been used in organization theory and social sciences (Gabriel, 2000; Czarniawska, 1998). Today, narrative studies are being established as a legitimate form of research methodology and have been employed in a variety of settings and research projects (see, e.g., Pentland, 1999). Gabriel (2004: 2) points to the recognition of narratives and stories as valuable empirical material in organization theory and management studies:

> Long tarnished as mere hearsay, opinion, or invention, stories, with all their inaccuracies, exaggerations, omissions, and liberties, are now seen as providing vital clues not only into what happened, but what people experience, or even into what they want to believe as having actually happened.

What was previously ignored as what was human, all too human, and dependent upon individual worldviews and idiosyncratic thinking is today treated as what provides valuable inroads into the collective production of meaning in organizations (Bruner, 1986; Polkinghorne, 1988). Gabriel (2004: 20) emphasizes that the value of a narrative or story may not lie in an 'accurate depiction of facts' but in 'its meaning', in its ability to provide a shared ground among communities of human beings such as co-workers in an organizations. Cunliffe, Luhman and Boje (2004: 276) also point to the close associations between narratives and meaning: 'Meaning unfolds in narrative performance, in time and context, as storytellers and listeners discuss their experiences, inter-weave their own narratives: a polyphony of competing narrative voices and stories told by many voices within different historical, cultural, and relational contexts.' Since human beings formulate narratives and tell stories about how they engage with practical undertakings and create sense and meaning in such *mimesis* of the real, narratives are meaning-bearing *social facts* (Durkheim, 1895/1938) and need to be interpreted and examined as such. However, this does not imply that narratives are to be examined at the surface of face validity but rather need to be regarded as a form of complex social interaction. For instance, in an interview situation, the interviewer may ask the interviewee to give an account of a specific event. The narrative that (hopefully) follows is not drawing on a pool of brute facts but is making use of a series of hetero-geneous materials that comes to mind in that specific situation. Shotter and Billig (1998: 16) write:

> [W]hen people talks of remembering in everyday life – when they make 'memory-claims' – they are rarely, if ever, simply describing or reporting an internal process or mental state: they are engaging in the rhetorical, and often contentious, activity of social life, and telling of, or expressing, something of their own position in the current scheme of things in relation to others around them.

Some researchers are taking the consequences of this view even further and claim that the very production of the self, of a coherent and intelligible subject, is inextricably bound up with a narrative capacity. Benhabib (2002: 15), for instance, argues:

> To be and to become a self is to insert oneself into webs of interlocution; it is to know how to answer when one is addressed and to know how to address others...Strictly speaking, we never really insert ourselves, but are rather thrown into these webs of interlocution, in the Heideggerian sense of 'thrownness' as *Geworfenheit*: We are born into webs of interlocution or narrative, from familiar and gender narratives to linguistic ones and to the macronarratives of collective identity. We become aware of who we are by learning to become conversation partners in these narratives.

In Benhabib's view, we become what we are or want to be as a consequence of what personal or social narratives we draw on (see also Wajcman and Martin, 2002; Humphreys and Brown, 2002; de Peuter, 1998). Social identities and the use of shared language games and forms of communication are therefore tied to one another. However, even if one should not underrate the collective production of social conditions through the use of narratives, narratives also play the role of being not only a 'mode of communication' but also a 'mode of knowing' (Czarniawska, 2004: 6). Even personal knowledge that you do not even share with others tend to become structured – *emplotted* in the favoured vocabulary – around certain sequences of events, time frames, or rules of engagement (Patriotta, 2003). As a consequence, narratives are not only vehicles for sharing and jointly constituting knowledge, meaning and understanding (Currie and Brown, 2003; Orr, 1996; Donnellon, 1996; Boyce, 1995; Knorr Cetina, 1995; Boden, 1994), but is also part of what Polanyi (1958) calls *personal knowledge*, and after Polanyi has been called *tacit knowledge* (Gourlay, 2004; Tsoukas, 2003).

In summary, then, narratives are increasingly recognized as important social facts that both practicing managers and researchers need to understand and exploit in the day-to-day work in organizations and in research projects. Knowledge-intensive organizations (Alvesson, 2001; Starbuck, 1992) are highly dependent upon the ability of its co-workers to jointly share and create knowledge and engage in creative solutions to practical problems and therefore the management of organizational creativity needs to underline the integrative, yet explorative qualities in narratives.

In the following, we will offer a number of examples of how narratives can address issues and concerns in creative work.

The fallacy of misplaced concreteness and the management of organizational creativity

Many concepts in organizational and managerial literature have an underlying problematic issue that demands attention – namely, their ambiguous nature and the way in which they are treated in an organization. In science and in our everyday thinking, we use ideas about the world, which we have simplified and abstracted from our experience. Of course, this is convenient, and even necessary, for many purposes, and no harm is done as long as we remember that we *are* simplifying and abstracting. The danger appears when we forget this and mistake the abstract for the concrete, creating problems for ourselves (Burke, 2000). For example, leadership, empowerment, diversity and creativity are concepts we employ to create stability, form, and predictability, but they are ambiguous, and there is a risk that they will fall under what Whitehead (1927) calls the 'fallacy of misplaced concreteness'.

This error, Whitehead explains, is to 'neglect the degree of abstraction involved when an actual entity is considered merely so far as it exemplifies certain categories of thought' (Whitehead, 1925: 51). Or, as Young (1994: 75) comments, 'whereby abstractions are created for a quite distinct purpose, but that purpose is forgotten and one becomes stuck with them and equates them with reality, substituting them for direct experience, which comes to be experienced in terms of the equation between that set of abstractions and reality itself'. The underlying problem is the *notion of simple location*. For Whitehead, being a process thinker means rejecting the notion of simple location; there are no entities, merely events, becoming moments. Whitehead's philosophy formulates a vision of reality, not as 'here, now, immediate, and discrete' (Whitehead, 1933: 180), but as an ambiguous and unfinished process. Whitehead's philosophy can be defined as the philosophy of organism: open-ended and taking into account – apart from scientific materialism – that all aspects are an ongoing interrelated process, and that the process is the ultimate reality.

The doctrine of simple location, or the logic of presence, follows from the atomistic view of placing things at simple points (sometimes space-less and time-less), which assumes that the opportunity exists for the attainment of full and immediate meaning and presence through concepts and terms (Chia, 1998). So the notion of simple location does not take into account that events are always on the move, because they constantly emanate from their place of origin. In so doing, they are governed neither by the confinement property nor the externality property. In other words, it means that nothing is isolated, simply being complete in itself. According to Chia (1995: 590), Whitehead claims that the fallacy of misplaced concreteness is a result of 'the modernist tendency to reify, invert and forget and thereby view the world as being made up of a succession of discrete configurations of matter (i.e., individuals and organizations)'.

Whitehead, for example, asks in what sense is its scent the property of the rose. He points out that in the absence of noses there would be no smells. Likewise without eyes there would be no colours, and without ears, no sounds. These are all *constituted* qualities and it is the organization, ordering, and abstracting of certain sensations – seeing, hearing, touching, etc. – that constitutes the order of nature:

> These sensations are projected by the mind so as to clothe appro-
> priate bodies in external nature. Thus the bodies are perceived as
> with qualities which in reality do not belong to them, qualities
> which in fact are purely the offspring of the mind. Thus nature gets
> credit which should in truth be reserved for ourselves: the rose for its
> scent: the nightingale for its song: and the sun for its radiance. The
> poets are entirely mistaken. They should address their lyrics to them-
> selves, and should turn them into odes of self-congratulation on the
> excellence of the human mind. Nature is a dull affair, soundless,
> scentless, colourless; merely the hurrying of material, endlessly,
> meaningless. (Whitehead, 1925: 54)

Whitehead thus represents the idea of the open system as one in which entities – people, ideas, and things – require each other for their existence. An open system is one that exhibits dynamic ambiguity, self-organization and unpredictability (Rose, 2002). In other words, Whitehead wants to unmask the scientific and philosophical desire to abstract and fix reality in a singular expression. He wants to convey a suspicion about the over-intellectualization of reality to prefer that different inter-pretative methods co-exist. Or, as Sherburne (1966: 195) points out 'The aim of generalization is sound, but the estimate of success is exaggerated'. Reality, in Whitehead's view, is a fluid interplay of relations between concrete actualities and infinite possibilities.

At a more general level, the fallacy of misplaced concreteness can be seen as a kind of belief that turns into an unnoticed or unconscious delusion, which then leads to unwarranted conclusions about concrete actuality (Daly and Cobb, 1990). In Whitehead's view, therefore, it is a fallacy to think of for example that employees, and leaders, as being simply located – here, now, immediate, enduring, and discrete – without any reference to prior and following events. A relevant example is that an individual in an organization is not creative *per se* (simple location); it is her or his ability to interact in a whole assemblage of organizational entities (e.g., sharing information, networking, communicating, connectivity with others, generating ideas, and receiving acceptance for ideas) that makes the person creative. Thus, creativity is *distributed within a place* (i.e., an organization). Organizational creativity can be seen as a 'systematic complex of mutual relatedness' (Whitehead, 1925: 161), that is to say actors (employees and managers) is affirmed in becoming – and not a property of a self-determining individual.

Ambiguous concepts, as mentioned above, may thus end in oversimplification and ambiguity, which may become meaningless for the organization; there is a link to the context of the concept and related processes, and the individuals behind it. In addition, over-ambitious use of metrics in companies is also within the fallacy risk zone. This leads to the important task of defining and describing (1) what organizational creativity is and (2) what it means for the organization—trying to clarify and to predict possible outcomes, effects, and cost (in concrete terms), while being sensitive to the need for highlighting and clarifying causal relationships of ambiguous concepts by breaking down different concrete effects that are relevant to organizational reality. But if the concept is treated in an abstract way, neglecting to identify causal relationships and effects, and not carefully checking the costs of the organization, then it may turn out that promising initiatives simply stall because the effects are not in balance with what the organization can afford to change (Sundgren, 2003).

The liminality of creativity

Following Alfred North Whitehead, we here conceive of organizational creativity in processual terms, that is, something that cannot be simply located in single points or entities but which is distributed and shared. Nevertheless, recognizing the process-based view of organizational creativity does not imply a smooth succession of states into a series of creative accomplishments, but the process-based view also acknowledges the qualitative leaps between different points wherein certain scientific findings are recognized and constituted *qua* scientific facts. In between such 'points of recognition', scientific work is of necessity located in what we in the following section will refer to as a *liminal domain* characterized by the absence of fixed rules and pre-established facts. The movement from hypothesis to established 'fact' (Latour, 1987) therefore implies the passing through a liminal phase where there is no certainty or agreement over the implications of the scientific results or findings. The notion of liminality (from Latin *limen*, 'threshold') is used in anthropology literature and denotes transient phases in the course of life wherein the individual is temporary released from the regular institutions only to be re-integrated in the next phase. Common liminal phases in tribal societies are passage rites wherein, for instance, a child is being confined and excluded from society in order to pass a number of tests that makes the individual qualify as a legitimate adult of the particular society. In our modern societies, there are still a number of such passage rites in use: the communion among Catholics and the Bar Mitzwah among Jews are two such examples. In this section, organizational creativity will be examined in terms of being located in liminal domains. Prior to this discussion, anthropological theory on liminality will be examined. Although the notion of liminality was developed in anthropology, it is used in a number of studies of organizations and management practice. Czarniawska and Mazza (2003) discuss consultants as being a form of liminal subjects being located in between the organization and the market. Garsten (1999) examines temporary workers (so-called 'temps') as liminal subjects passing between different organizations. Tempest and Starkey (2004) explore how organization learning is affected by individuals' liminal positions within firms.

The notion of passage rites and liminality was first coined and examined by the anthropologist Arnold Van Gennep (1960). Van Gennep's terminology was later used by Victor Turner in a series of studies of tribal societies. For Turner, the liminal phase is of necessity ambiguous because it denotes a position that is external to the regular social organization. Turner argues that the liminal subject is no longer firmly located within the social texture but is rather 'betwixt and between' social positions (e.g., childhood and the status as adult):

> The attributes of liminality or of *liminal personae* ('threshold people') are of necessity ambiguous, since this condition and these persons elude or slip through the network of classifications that normally locates states and positions in cultural space. Liminal entities are

neither here nor there; they are betwixt and between the positions assigned and arrayed by law, custom, convention, and ceremonial. (Turner, 1969: 81)

For the *liminal subject*, the individual located in liminality, a social role is abandoned and replaced by a position outside of a particular social order. In order to deal with these ambiguities and temporary loss of identity, Turner argues, new coalitions are constructed among individuals sharing the same predicament of being liminal subjects. Such coalitions are named *communitas* by Turner (1969). For Turner, not only phases of life may be regarded as liminal phases but entire societies have mechanisms that enable the recognition of ambiguities inherent to liminality. For instance, the medieval carnival (Bakhtin, 1968; Girard, 1977) – still present in many late modern societies in the form of festivals – turning society upside down in order to reproduce the social order, is one significant example of liminality. 'The liminal phases of tribal society invest but do not usually subvert the *Status Quo*, the structural form, of society', Turner writes (1982: 41). For Turner, such combinations of liminality and the structured form of society can be examined in functionalist terms as what makes society durable and capable of dealing with ambiguities. Turner (1969: 193) writes:

Society (societas) seems to be a process rather than a thing – a dialectical process with successive phases of structure and communitas. There would seem to be – if one can use such a controversial term – a human 'need' to participate in both modalities. Persons starved of one of their functional day-to-day activities seek it in the ritual liminality. The structurally inferior aspire to symbolic structural superiority in ritual; the structurally superior aspire to symbolic communitas and undergo penance to achieve it.

Each society – ancient, tribal, modern – has established its idiosyncratic forms of liminal phases and liminal institutions. In our society, the teenage years or the years as college or university student may be treated as instances of liminality (and a certain degree of subversion) within an otherwise well-structured society.

Speaking in terms of organizational creativity, the notion of liminality is helpful when exploring the experiences of being 'betwixt and between' a solution to a complex scientific – or non-scientific, for that matter – problem. Numerous accounts of scientific breakthroughs contains stories of how individual researchers are dedicating substantial time and effort to specific problems without knowing when and how to solve the problem but only that they cannot abandon the problem at that point of time. Scientific breakthrough – by definition creative events – are often preceded by substantial efforts. Being on the verge of solving a problem is of necessity to be located in a liminal position; you cannot fully draw on previous experiences and know-how (even though such resources are a *sine qua non* for creative work), yet you cannot be sure what the outcome of your efforts will be. You are neither here not there but betwixt and between what is already known and what may eventually be regarded a fruitful contribution. While the regular social structure – for instance, the standard operating procedures and routines of an organization or a laboratory – is helpful in monitoring and making use of already existing knowledge, what has already been qualified, judged and agreed upon as legitimate, 'factual' knowledge, the *knowledge-in-the-making* can never be constituted solely on basis of such structures. Instead, the knowledge-in-the-making needs to effectively move beyond what is already at hand in order to apprehend what is new. Creative work is therefore uncertain, demanding, fraught with ambiguities, and, of necessity, implies numerous setbacks, disappointments, cul-de-sacs, and failures. Behind the grand modernist narrative of scientific progress and the mastery of nature, there is a hidden and rarely recognized history of scientific efforts that led nowhere.[34] Speaking with Max Weber (1948), the scientist needs to be equipped with not only a long time perspective and the awareness that any scientific achievements is doomed to become obsolete and forgotten, but also the insight that any contribution to a field is, of necessity, small and narrowly defined.

Organizational creativity is what is taking place within this liminal domain wherein human faculties such as intuition are helpful resources enabling the identification and detection of significant patterns of what may later be brought to the test of proof. Anecdotal evidence suggests that researchers in the pharmaceutical industry do only occasionally speak of organizational creativity, even though they recognize the importance of it for their work and the vitality of the industry. While the regular day-to-day management of laboratory work is captured by a detailed vocabulary and terminology, creativity appears to defy such signifying frameworks. Since management as ideology and practice is

---

[34] In Bowker's (1995: 583) formulation: 'History is a success story'; 'For every book about the dead-ends, the failed experiments, the frauds, there are thousands about Kepler, Newton or Einstein' (Bowker, 1995: 583).

rather poorly developed in terms of addressing what is process-based and ambiguous, the notion of creativity is little used, albeit being of significant importance for the activities. Such failure to speak of what is occurring in the liminal domain is what is threatening to the organization's sustainable competitive advantage. Therefore, the management of organizational creativity implies a recognition and an awareness of the ambiguities, fluidity and complexities of the creative process.

The political economy of creativity

Organizational creativity is an important resource or capability underlying sustainable competitive advantage for organizations. At the same time, organizational creativity implies certain trade-offs, investments, negotiations, and other forms of political decisions regarding the use of resources. Organizational creativity does not, one may argue against the views of popular organizational creativity writers, come without attachments. On the contrary, the whole idea of organizational creativity is embedded in the highly political setting of the organization. As have been emphasized in a number of studies of decision-making (March and Olsen, 1976; Pettigrew, 1973; Allison, 1971), decisions are always based on a variety of choices, opportunities and objectives that are advocated by different stakeholders. This makes decisions a highly politicized event in organizations, and because organizational creativity is always dependent upon decision-making (e.g., the decision what new chemical entity among a number of choices to elaborate upon in pharmaceutical industry) one may speak of a *political economy of organizational creativity*. The notion of political economy is invoked in social science literature and philosophy to denote trade-offs and choices between alternatives. However, the concept has changed in meaning over time. Marx and the economists of his time talked about political economy when we today are prone to say 'economics'. Today, it is used when pointing at certain trade-offs. Foucault (1980: 131) is speaking about the '"political economy" of truth', suggesting that the idea of truth is by no means a transcendental and absolute category but what is negotiated and agreed upon within discursive formations. The French urbanist and architect Paul Virilio, examining the increased speed in the contemporary society, talks of the notion of 'political economy of speed' (Virilio, cited in Armitage, 2001: 161). Korczynski (2000), discussing the importance of trust in organizations, similarly advances the idea of a 'political economy of trust' in organizations, and Noam Chomsky has used the expression 'the political economy of human rights' to capture the overtly pragmatic attitude in terms of privileging economic interests towards what are supposed to serve as universal and immutable laws. The notion of the political economy of scarce resources is therefore widely recognized and employed in a variety of discourses. As a consequence, it is possible to speak of a political economy of organizational creativity when stressing creativity's embedding in the access to scarce resources. However, this political economy of creativity is primarily oriented towards the social components of creativity, what we in Chapter 3 called the 'layers' of creativity; a significant part of new drug development research – what we previously called 'the core' – is determined by scientific paradigms and their instituted lines of demarcation between truth and non-truth, the legitimate and the non-legitimate contribution. What remains outside scientific interests are subject to such political economies of creativity – that is, choices and decisions among competing alternatives. Navigating in a political field does not only require skills and experiences but also an ability to, in Goffman's (1959) terms, 'present oneself' in the correct manner (Naurin, 2004). One may therefore speak of 'political entrepreneurship' as what is purposefully and deliberately exploiting the resources available within the political economy of organizational creativity. The political entrepreneur is then an individual capable of advocating his or her ideas within a field of competing choices and objectives characterized by scarce resources. In the following, the notion of political entrepreneurship will be examined.

Political entrepreneurship and creativity

It is very common in creativity and management literature to parade the successful entrepreneur, inventor, or innovative company in order to celebrate the benefits of creativity. This celebration of the victor is only one side of the coin. The other side is what can be called the *political economy of creativity* in organizations: management can make a decision about 'how much creativity' they believe satisfies the need for the organization to renew its product or service portfolio. The investment in creative activities will always produce certain types of organizational activities – the consequences of which may not always be desirable. Moreover, organizational creativity implies, by definition, a deviation in some sense from the standardized way of doing things, which includes, for example, persistence, flexibility, and opposition. Project groups may reformulate problems and objectives when facing problems, rather than continuing down the same path. Because creative processes are always non-linear and disruptive, and are based on the interaction of tight and loose systems, creativity is costly and demanding of resources that must be managed and controlled. Although creativity is not a good thing per se, it can also be detrimental to organizational activities in cases where stability, predictability, and manageability are highly needed and praised. Creative activities are at the very heart of organizational renewal, but it may be that it is misplaced at times. In summary: *organizational creativity begets the new* (Whitehead, 1927). As such, it always challenges the existing culture and power structures (Staw, 1995); structures that are difficult to change.

There is an uneasiness to confront the reality that organizations are not in fact stably ordered, predictable, rule-based systems, or, as Buchanan and Badham (1999: 1) put it: 'We perhaps like to think of our social and organizational cultures as characterized by order, rationality, openness, collaboration and trust. The reality is different'. Thus, political behaviour plays a more significant role in organizational life than is commonly recognized. Buchanan and Badham (1999) argue that the role of political behaviour in organizational change processes has been only occasionally discussed in the academic management literature. Contrasting the rational actor model, Dill and Pearson (1984: 139) argue that a model acknowledging organizational politics is advantageous because it perceives 'the pluralistic needs and values of organizational participants, sources of individual power, the importance of informal communications networks and the consequences of these factors for defining the necessary managerial skills'. Political behaviour is connected to power. Harold Lasswell's (1950) definition of politics – politics is who gets what, when and how – is among the more famous. Buchanan and Badham (1999) offer some useful definitions in this area. *Power* can be seen as the capacity of individuals to exert their will over others, and, *political behaviour* is the practical domain of power in action, worked out through the use of techniques of influence. Thus political behaviour can be seen as activities and behaviors in order to get things done 'your way' (Astley and Sachdeva, 1984) or, as Euston (1965) defined political life, as a set or system of interactions defined by the fact that they are more or less directly related to the authoritative allocations of values for a society. The notion of entrepreneurship is often viewed as a function or ability that involves the exploitation of opportunities which exist within a market, or an organization. In particular, entrepreneurship can be seen as a combination of the exploration of opportunities and the bearing of uncertainty (e.g., Kirzner, 1973; Knight, 1971). Political entrepreneurship (e.g. Kingdon, 1984; Schneider and Teske, 1992) represents a kind of presence of individuals in organizations who are committed to the cause.

Political entrepreneurship requires the ability to operate in an organization, combining a flexible number of skills with enabling activities such as intervention in political processes, pushing particular agendas, rational persuasion, influencing decisions and decision makers, dealing with criticism and challenge, coping with resistance, and promoting credibility in order to reach objectives or goals (Buchanan and Badham, 1999). For example, a project leader needs to have a pre-understanding of the organization's power structures and politics. This enables the project to conform with the political conditions prevailing without compromising the project. Kakabadse (1984) has proposed six useful guidelines for the political entrepreneur: *identify the stakeholders; work on the comfort zones; network; make deals; withhold*; and *withdraw*. On a generic level, political entrepreneurship can be defined *'as the exploitation of opportunities in order to allocate scarce resources to outcomes and preferences'* (Björkman and Sundgren, 2005).

The creative process in organizations requires understanding, patience, and an awareness of organizational politics. Thus, the ability to manage political relationships is essential for influencing and ensuring the legitimacy of new ideas and for increase the possibility to get a fair evaluation of creative ideas. Despite the surprising absence and negligence of the political dimension in creativity literature, the concept of political entrepreneurship has great relevance and importance for managing organizational creativity. For employees (researchers and managers), political entrepreneurship may, for example, mean success or failure in ensuring the adoption and diffusion of an innovation (Frost and Egri, 1991). So creativity is a potential key to success and a capacity that can be used to pursue political means within the organization (Styhre and Sundgren, 2003b). On a practical level, political entrepreneurship may involve creating attention for new ideas, facilitating support and funding for testing new concepts, or using rational persuasion on initiating new projects, in short, knowing the limits of the politics of creativity and pushing them. To summarize, managing creativity always includes trade-offs between stability and renewal, predictability and emerging opportunities, and efficiency and entrepreneurialism: a political economy of creativity. Political entrepreneurship is important for operating in this economy.

Managing organizational creativity

While the first four sections of this chapter have discussed a number of aspects pertaining to the management of organizational creativity – the creativity developed and exploited within organizations – there have been only a modest number of suggestions as to how organizational creativity can actually be managed in more practical terms. In the next two sections, a number of detailed and practical suggestions drawing on the themes addressed in the book are discussed. There is no suggestion that these are universally applicable or flawless; rather, they are based on observations and insights gained from the companies examined in the book. There are, of course, limitations and industry idiosyncrasies that need to be examined but the suggestions are here formulated in general terms and are therefore, in our view, adequate for a variety of activities and industries. In this section, we make a distinction between *practices*, activities that can be undertaken to support and nourish organizational creativity and *worldviews*, cognitive and attitudinal aspects of organizational creativity.

Practical considerations

1 Managing the political agenda

Popular myth holds that 'you cannot hold a creative idea back': Sooner or later a good idea will always be recognized and treated with the attention it deserves. Such social Darwinist beliefs may hold for longer periods of time, but are conspicuously ignorant of the inertia demonstrated in society (see, e.g., Bourdieu and Passeron, 1977) and also speak of entire societies rather than specific organizations. A new and creative idea is always a Trojan horse; it brings new thinking into a domain that is potentially hostile because it is not capable of dealing with what is new and breaks with the received wisdom. Empirical research in new drug development shows that what proved to be highly successful drugs when reaching the marketplace were often initially treated with scepticism. In fact, out of seven blockbuster drugs (drugs rendering more than US$1 billion in annual income) studied, all new candidate drugs were terminated at least twice during before finally making it to the market. Therefore, organizational creativity does not happen on its own but needs to be rooted in joint agreements on those kind of ideas and findings which are worthwhile and which support the firm's strategy. In other words, leaders and managers need to *advocate* and *actively support* those creative ideas which they believe hold the greatest potential.

Many creative ideas – some would say this may serve as a fruitful operating definition of creativity – are born into a hostile environment or at
least an environment where different ideas are competing for scarce
resources and therefore the political dimension of organizational creativity needs to be taken into account.

2 Develop leadership skills and practices

Creative work is highly demanding both for the individuals engaged in
problem-solving but also for those in charge of the project. Creative
processes tend to be non-linear, unpredictable, and emerge in a disruptive
manner. On the other hand, the management of operations demands
transparency, objective measures, and predictable results. One of the
key arguments of this book is that organizational creativity is not a
self-organizing process; individual co-workers may always come up
with creative ideas and interesting thoughts, but a social system such as
a company need to be managed in order to demonstrate organizational
creativity over time. In many cases, the very idea of management is
treated as what is aimed at determining and fixing what is ambiguous
and fluid; management as practice is then of necessity what is incapable
of dealing with creative processes. However, the notion of management
can be filled with other meanings. For instance, management can
equally be the capability to support and nourish thinking and activities
that demand an acceptance for non-linearity and predictability. Therefore,
leadership skills and practices need to be developed and continuously
elaborated upon and discussed in companies that believe they are
benefiting from organizational creativity. For instance, a self-reflective
attitude towards leadership work and management practice may be of
great help in creative environments: 'Are we capable of monitoring
creative work?', 'What kind of support can managers offer creative
co-workers?', and 'How do we improve leadership skills?', are questions
that need to be considered. Leadership is important and it can be just as
helpful as it can effectively eliminate a creative environment.

3 Measuring and monitoring creativity

Organizational creativity is a frail construct in terms of, in many cases,
escaping the conventional managerial practices such as accounting and
other control systems. Still, organizational creativity can be measured
and monitored by instruments that provides, speaking with Charles S.
Peirce (1991), *indexes* for creativity. Peirce distinguishes three forms of
signs: *Icons* are direct illustrations of underlying realities e.g., the waste

bin on the computer desktop serving as an illustration of a real waste bin; *indexes* that are pointing at a causal relationship, e.g., a knife and fork on a sign representing a restaurant, and, finally, *symbols* that are arbitrary, e.g., the Union Jack representing the UK or a logo representing a company (e.g., Nike's 'Swoosh' logo). Measurement of creativity may not capture creativity per se but may instead point to the precursors for creativity. For instance, the notion of *creative climate* advocated by Ekvall (1999, 1982) points to a number of social, behavioural and cultural conditions that can predict creativity. In Peirce's terms, such a creative climate is an index of organizational creativity. The management of organizational creativity implies the use of such measurements and monitoring practices, encircling organizational creativity per se, but still of great value when predicting creative outputs. Even though organizational creativity has been portrayed as what is ambiguous and fickle, this does not suggest that conventional management practices are impotent. Instead, such practices need to be adapted carefully to reflect local conditions and interests. The measuring and monitoring of factors serving as precursors to organizational creativity is especially helpful when being connected to incentive systems and extrinsic motivation. Although creativity is essentially drawing on intrinsic motivation, such innate 'callings' (Weber, 1948) may be supported by carefully designed incentives and reward systems. Similarly to the discussion about the use of technology, it is important to keep in mind that measurement and monitoring practices does not displace organizational creativity per se but should rather serve as the infrastructure (Bowker and Star, 1999) of the creative organization.

4 Managing technology as a means, not an end

Technology is one of the most significant human achievements and has raised the standard of living to unprecedented levels. Technology is not only being used as a tool in various undertakings, but is today also starting to penetrate the human body (e.g., nanotechnology in medical treatment) and is therefore wielding an even larger influence on human beings' lives. However, technology does not speak for itself and needs to be carefully mastered and monitored by humans in order to function properly. In terms of organizational creativity, technology is, in many cases (and especially in the laboratory sciences serving as the backbone of pharmaceutical industry) an indispensable resource. Nevertheless, investment in technology needs to be accompanied by the full support and recognition among professional groups and communities of practice subject to changes in work practices. Technology thus needs to be carefully embedded in predominant scientific ideologies and practices. Expressed differently, technology needs to be regarded a means, not an end in itself; technology does not substitute for organizational creativity, but can only serve it as one resource among others in creative work. The interview material in Chapter 7 suggests that laboratory scientists share a general scepticism towards technologies such as high-throughput screening because it is thought of as degrading laboratory work. Other industries may be capable of demonstrating similar attitudes. Therefore, technology should be seen as a means rather than an end in creative work.

5 Narrating organizational creativity

An essential point in the construct organizational creativity is that it is what is emerging and is exploited within a social organization: a firm, a community, a team, or a group. In order to make what is inherently ambiguous and fluid intelligible and to make it possible to share it with others, humans engage in storytelling. In narrating experiences, insights, and know-how, individuals are trained in becoming able to share their experiences. In many cases, there is little discussion or debate regarding creative work *qua* creativity; creativity is not a standing issue in conversations but is, on the contrary, very much excluded from the day-to-day discussions. This is not just an anecdotal remark, but is actually of great importance for the firm's ability to exploit its intellectual capital. Without a shared linguistic framework, researchers and other creative groups are incapable of making their activities transparent to others. One of the consequences is that organizational creativity needs to become a part of the day-to-day agenda and what is explicitly encouraged to bring up as a topic of discussion.

Worldviews: Perspectives on organizational creativity

6 Managing intuition and cognition

As was discussed in Chapter 6, organizational creativity may draw on what is escaping the vocabulary of a rationality that assumes that all thinking can be given a proper expression – that language mirrors thinking. Following Henri Bergson, intuition is what is appearing in-between signifying systems such as scientific languages and therefore one cannot expect creative co-workers to be capable of fully accounting for every single step in their course of progress. In fact, during some stages of creative work, researchers and other creative individuals are operating without a proper language systems capturing what is in a state of becoming. From a managerial point of view, such inabilities may appear as a form of ignorance or incompetence, but at these stages, in Whitehead's (1938: 49) apt expression, 'language halts behind intuition'; language has not yet been developed and agreed upon and the researchers cannot afford to take the time to develop it because they are dealing with other concerns and therefore they develop their own provisional and local language-games (see MacKenzie, 1999). Later on, when the most intense creative stages are coming to an end, researchers are likely to dedicate more time to the formal language denoting their output. The leadership and management of organizational creativity must be capable of enduring such stages of ambiguities. Intuitive thinking and other forms of 'competing rationalities' is therefore a pivotal organizational resource and need to managed as such.

7 Maintaining a process-based view of creativity

In Whitehead's (1925) account, the human intellect tends to disrupt what is fluid and processual and break it down into discrete categories and fixed entities. This fallacy of misplaced concreteness is not, one may argue, an esoteric philosophical reflection removed from everyday work life practice, but is rather a reminder that much management and leadership practice is operating on the basis of such reductionist treatments; organizations are, by definition, what is structuring and ordering a variety of activities and entities into a coherent and transparent whole that can be easily monitored. In that respect, all organizations are what Jeremy Bentham (1995) calls *panopticons*, a form of social organization aimed at maximizing visibility at the lowest possible costs and efforts. In terms of organizational creativity, the practices of, say, laboratory work are organized into workspaces, work groups, work positions, routines, standard operating procedures, and other such matters. This organization is the infrastructure for all creative work, vital for the long-term capacity of undertaking creative work, but it is easy to mistake the means for the ends and to regard the organization per se as what is producing creative solutions and findings. Just because the organization apparatus is structured in a sequential manner into a series of operations, this does not suggest that creative work per se unfolds in accordance with this particular form of organizing. Organizational creativity may at times be best conceived of in non-linear terms while, in other cases, it may unfold in a more straightforward and sequential manner. Thinking of organizational creativity in more processual terms therefore puts pressure on practicing managers to think outside the specific organizational model they are in charge of. Managing organizational creativity is in other words a matter of cognition, of how the world is apprehended by organizational actors.

8 Recognizing the influence of liminality

Organizational creativity is by no means an organizational capability that is easy to manage and monitor; it appears at unpredictable points in time, its long-term consequences are not easily estimated, and it does not conform to the regular tool-box of the practicing manager. In addition, communities that are capable of presenting creative solutions to defined problems do not always respond to the forms of extrinsic motivation (e.g., rewards, bonuses, salaries) that have traditionally been developed in organizations; rather, they are driven by intrinsic motivation and credibility that certain breakthroughs generates within certain communities (Latour and Woolgar, 1979). Moreover, the creative work always of necessity implies some kind of liminal experience, that is, being located in unknown terrains wherein one is in the state of solving a specific problem. Since the management tools and practices at hand generally do not deal effectively with such liminal experiences, the management of organizational creativity demands a certain degree of tolerance and patience with creative communities. One cannot eliminate the liminal phases in the creative sequence; rather, they must recognize the need for political and emotional support during such periods of being betwixt and between a solution, or, at least, the temporal establishment of a new state of the art. Managing organizational creativity therefore includes the ability to endure uncertainty and ambiguities and the need to provide the required political, emotional, and – if possible – intellectual support for the individuals engaging in creative work should not be underrated.

Summary and conclusion

The first part of this book criticized the literature on organizational creativity for failing to address ontological and epistemological issues of organizational creativity. In the second part, we aimed to discuss a number of concerns regarding the management of organizational creativity that generally are underrated, marginalized, or ignored in the literature. Organizational creativity is not a resource or a capability that is easily captured by brief definitions and neither is the management of such resources. Therefore, the recommendations on 'how to manage' need to be approached with a degree of scepticism. It is not our intention to provide conclusive arguments and checklists but rather to point to some of the issues that would be helpful to place on the agenda when making claims to be an organization or company that is drawing on its capacity to exploit organizational creativity. Even though we are critical of the belief in capturing organizational creativity by one single unified model and a set of attached practices, we believe that it is possible to learn from others and to identify needs and demands that are relevant for many industries and companies. Therefore, we think that a consideration of the seven topics discussed above is of relevance for any organization making claims to be creative.

The future of pharmaceutical industry and creative work

In recent years, the pharmaceutical industry has undergone radical transformation and this has resulted in the formation of large, merged, global companies in order to optimize the rapid worldwide launch of products, to ensure shared R&D infrastructure costs, and to minimize product risk by sustaining a broader project portfolio in which strategies have been formed by the resource allocation decisions made during the early R&D phases, the competitiveness defined by the R&D pipeline, and the innovation process embedded into a complex governance logic derived from the stock market. Studies show that the R&D productivity of pharmaceutical firms, measured as the number of new medical entities registered on the major markets versus overall R&D investment, has decreased significantly during recent decades (Taylor, 2003). Fortunately, many large pharmaceutical companies seem to sustain a highly viable business by focusing on the development and production of a few very large *mega brands*. Successful companies, surpassing decades of challenge, have retained strong company values, which include a commitment to R&D based on internal innovation.

Together with increasing competition in the pharmaceutical marketplace, which has resulted in an increased focus on blockbuster drugs, and on a stronger regulatory environment, this transformation has essentially reduced the degree of freedom in pharmaceutical practice and resulted in much larger, more complex R&D organizations (e.g., Schmid and Smith, 2002b; Dollery, 1999). Among several consequences for the industry, large multinational firms tend to become increasingly process-oriented and to rely on standard operating procedures and other forms of bureaucratic routine and standardization (Hullman, 2000). According to some critics, a good start is to forget 'me-too market'[35] (after compounds three and four

---

[35] 'Me-too market' or 'Me-too drugs' refers to the idea of taking a share of already established markets by producing a drug that is something similar to a top-selling drug (Angel, 2004).

are licensed) and go and find those receptors, old and new, and the genu-
inely new compounds to interact with them.[36] For some companies, such
as Merck, which has resisted mergers in the past, the overwhelming urge
may be to grow bigger still through such unions, where a better approach
might be to spin out some or all of their R&D to realize its full value, rather
than producing more on and failing to make it pay.

Yet another important aspect of the industry is that new drug development is not only an expensive but also a risky business. During the entire
research process, including the post-market launch, the project, or product,
can be terminated or withdrawn at any time because of new knowledge
that shows unexpected effects of the drug. One of the most notable
examples in this sense is the case when Merck had to immediately withdraw its blockbuster drug Vioxx (Rofecoxib), in September 2004. The
product had sales of $2.5 billion and accounted for 11 per cent of the
company's sales in 2004.[37] According to some estimations show that
despite a loss of revenue from the withdrawal of the product, Merck's legal
liability cost from Vioxx-related suits may become as much as $18 billion
(Oberholzer-Gee and Inamdar, 2004). Moreover, the trend is that pharmaceutical companies are now facing increased scrutiny from regulatory
authorities such as the FDA (which in the case of the US have also included
congressional hearings), which are becoming more suspicious with drug
safety as with approving new drugs. One example is when one of Pfizer's
biggest products, Celebrex (an arthritis drug) with 2003 sales of US$1.9
billion, in December 2004 had to suspend its advertisements for the
product while US regulators review new data that link the drug to an
elevated risk of heart attacks.[38]Another example is when AstraZeneca did
not receive approval for the investigational oral anticoagulant Exanta
(ximelagatran) because FDA considers safety aspects outweighs the anticoagulant's benefit.[39] Thus, the challenge sets for pharmaceutical companies
to become truly innovative have resulted in increased interest on how to
manage *organizational creativity* as an organizational capability within the
pharmaceutical industry (e.g., Thompson, 2001; Dollery, 1999).

To conclude, many large pharmaceutical R&D organizations are
being forced to become increasingly efficient in delivering projects,

---

[36]*Lancet* (2004), vol. 364, 25 September.

[37]*History News Network* (2005) 'The Vioxx Wake-Up Call', 17 January.

[38] *NewYork Times* (2004), 'Pfizer to Halt Its Advertising of Celebrex to
Consumers', 19 December.

[39] *www.fdaadvisorycommittee.com*.

products, and services. Daily control and monitoring of organizational activities has become more detailed and sophisticated, while there have been many attempts to empower employees and to implement new organizational routines and standard operating procedures in order to improve the firm's knowledge-based resources. In short, instant commercial returns are favoured over a long-term investment in creativity. So there are several trade-offs that must be taken into consideration in managing organizational creativity in the pharmaceutical industry. One is the balance of management control, efficiency, and other approaches to streamline the research process with more flexible ways of working. Another is to reconsider *new ways* of using internal resources, knowledge, and capabilities in a large R&D organization. But with the fast, demanding pace at which the industry is pursuing projects, the trade-off may be that the company is unaware of the need to create enough space for creativity, intuition, and radical questioning of what is already agreed upon, in short, for thinking that operates within the realm of new drug development. Organizational creativity in the pharmaceutical industry is a collaborative process that involves multiple domains of knowledge, and it is epistemologically complicated because it is an amalgamation of hard science-based truth and social construction in complex interaction with the market.

What are the alternatives for pharmaceutical companies to support and manage organizational creativity in the future? According to some critics and insiders is the industry forced to change (e.g. Drews, 2003) in order to survive. Can organizations be designed to encourage employees? Can traditional organizational structures and rigid hierarchies be abandoned without sacrificing productivity and accountability? Organizational creativity is about raising probabilities but the industry has become more influenced by instrumental rationality and subject to management control. Based on aspects presented in previous chapters, the following actions and propositions are discussed and might be considered in order to enhance probabilities for creative action and secure long-term investment in creativity.

Strategical and long-term considerations

1 Revising organizational structures

It is plausible that a company's creative potential increases rapidly with its size. The larger the company, the more likely that the components of creative act are already present, but less likely those they will be brought together without some help (Robinson and Stern, 1997). Moreover, a sustained high level of radical innovation rarely happens at leading large companies because of their inability to correctly value the future trajectories of market expectations and alternative core-underlying technologies (e.g. Christensen and Raynor, 2003). AstraZeneca, for example, has become a global company with a large, complex R&D organization and has undergone many changes, such as organizational size, technology, and various processes for improving research output. In comparison with a smaller pharmaceutical company, AstraZeneca has an enormous competitive advantage – if it can use the vast number of highly skilled employees, information databases, and financial resources to develop innovative products. But in this complex, planned, process-driven organization, creativity is a phenomenon that is harder to understand. Or, as one respondent expressed frustration regarding ways in which to communicate and distribute knowledge that is important for organizational creativity within AstraZeneca's large R&D organization:

> Scientific disciplines are becoming increasingly specific, complicated, and fragmented – the more you learn in a specific area, the less you understand. This is the feeling I have when going into an area. So today it's almost impossible to get an overview. Too few individuals are capable of grasping different disciplines: combining pharmacology and molecular biology, for example. So it's extremely important to create an organization with working and sensible interfaces between disciplines. These kinds of interfaces can create enormous drive and creativity in a project. This requires a flat organization, which we unfortunately are moving away from. (AstraZeneca, Senior researcher, Discovery)

Thus, there is an inevitable drawback, seen from an organizational creativity perspective, of being a large organization: information flows become more problematic, and silo thinking becomes an evident effect. Or, as Gordon Hewitt argues in the case of large pharmaceutical firms: 'as different types of innovations, both informatic and scientific, start to disrupt traditional assumptions about how the large pharmaceutical companies operate in a more value-driven and questioning environment...where traditional methods of management such as benchmarking and best practice have limitations in this kind of unpredictable environment. What is needed is to focus on what is "next practice" not "best practice".'[40] In this sense, there is clearly a need of organizatonal creativity at a high level to face the strategic challenge facing many large successful corporations. One practical example to solve this problem of having complex and bureaucratic organization is when GlaxoSmithKline fundamentally reorganized its R&D in 2001, splitting what the company believed to its more creative arm into smaller centres of excellence, and enlarging the arm that enjoys economies of scale.[41]

2 Creating new alliances, collaborations and joint ventures

The consolidation of the pharmaceutical industry, and the increased competition in the marketplace, increased demand on regulatory, and last but not least the brute economic power needed to take drug compound through Discovery, to Development and to the market has created large company structures and complex and bureaucratic R&D organizations (see Chapter 1). Therefore, focus on new collaborations is needed. Externally, the pharma industry needs to seek new collaborations, and joint ventures to link up with heterogeneous elements (biotech companies, regulators) to invest in new opportunities. Large pharmaceutical companies will have increasing problems if they aim to rely only on its own creative capabilities and live in splendid isolation. According to Hara (2003), the pharmaceutical industry in particular for radical innovation, or what he denotes as paradigmatic innovation such coherent organization may not be favourable because few members can accept the unfamiliar concept accompanying these types of innovation. Rather more heterogeneous organization with networks linking various external knowledge sources seem to be more appropriate for balanced innovation management in the pharmaceutical industry.

Moreover, pharmaceutical and biotechnology companies building their business on complex and interdependent technologies will face increasing problems in attempting to carry the necessary discoveries and development efforts on their own. Healthcare providers as well as funders will need to collaborate in developing the most effective treatment strategies. Hence, there is a need to find new collaborative approaches and connecting to major regional innovation systems, university and governmental research initiatives enabling more cost-effective

---

[40] Gordon Hewitt, Professor of International, Business and Corporate Strategy, at the Graduate Business School, University of Michigan. 'Pharmaceutical futures – reaping the full value of product lifecycle management', *AZ Source Magazine*, Issue 11, 2004.
[41] *The Economist* (2003), 'Big trouble for big Pharma', 6 December issue.

introductions of new treatments. Moreover, there is a growing need for an increased understanding of the driving forces for changing prerequisites, limiting assumptions and governing logics for pharmaceutical, and life science companies, health care providers and governments and the boundary-spanning nature of future activities and challenges will necessitate innovations in management models, systems and approaches.

Some pharmaceutical companies, most notably Eli Lilly and Pfizer, have started to outsource biotech development. Both companies reorganized internal research processes to make then more suited to research collaborations with smaller biotech firms. Eli Lilly has written a mission statement on partnering, while Pfizer has stated that one major corporate goal is to be the best partner in the pharmaceutical business. Among other developments, Eli Lilly has developed a secure e-mail 'bid-and-award' system for new projects sent to custom research organizations. Through a secure e-mail link between the companies, Eli Lilly provides the structure, process, safety information and literature references of the new project. The custom research organizations are then allowed to bid on the project. When completed, Lilly expects the system to allow purchasing to conduct the 'bid-and-award' process in as little as 24 to 48 hours, eliminating a lot of the legwork currently required.[42] Moreover, Eli Lilly also outsources custom manufacturing grammes to multi-kilogramme volumes, and intermediates to active pharmaceutical ingredients, process development or knowledge of how to make a particular molecule, analytical development, and new product development. In this sense their goal is to find a balance of internal resources and projects outsourced to custom research organizations. Eli Lilly argues that this experience makes the company a more attractive partner than its rivals for biotech firms and claim to have about 140 alliances with outside firms, both bringing in promising molecules and farming out its own. Today, each new product going into mid-stage clinical trials has a team of scientists, marketers and regulatory experts who work together to map out its future, from scrutiny at the FDA to patent expiry, ensuring that their molecule lives up to its full potential. This approach to managing product lifecycles may be more effective than the desperate machinations of some pharmaceutical companies to spin out patents on their own successful blockbusters.[43]

---

[42] *Purchasing Magazine* (2001), 'Dos and don'ts of discovery outsourcing', 5 April issue.

[43] *The Economist* (2002), 'Bloom and blight', 24 October issue.

3 New tools and culture for exploiting intellectual and information capital

Creativity and innovation are based on information, which makes it reasonable to expect that information sharing at all levels in an organization is important for the creation and successful transfer of corporate knowledge. The drive towards globalization has made it crucial for pharmaceutical firms to invest considerable resources in information infrastructures that can fulfill their varied information-processing and communication needs (Koretz and Lee, 1998). Information sharing is evolving into a technology of relationships, which facilitates the flow of interaction and associations through computer-based communication networks, groupware, increasingly intelligent agents, knowledge representation and management systems, databases, and convergence of different forms of traditional media (e.g., Cooper, 2003; Kao, 1996. 'Information', as used here, refers to a concept of strategic information that is related to the scientific and/or pharmaceutical projects and not primarily operative information related to cost reduction. There are expectations that increased information sharing will affect organizational creativity (Sundgren *et al.*, 2005a). In addition, in large multinational pharmaceutical companies access and reusing scientific data and information across project, therapeutic areas, functions and R&D sites has become increasingly more difficult (e.g. Pisano, 1997). Another factor in the pharmaceutical industries, at least compared to other industries, is becoming more information sensitive, which, to a large extent, relates to patents.

One example of how to improve the capability of reusing the organization's entire information capital is taken from AstraZeneca. The company's R&D organization had acknowledged that accessing and reusing scientific data and information across project, therapeutic areas, functions and R&D sites is becoming increasingly more difficult. In order to address this, a project called IM&KM (Information Management and Knowledge Management) was initiated in the global R&D organization. This project aims to link the global discovery and development organizations into a new framework on information and knowledge management across the new drug development process. Within this initiative, a common intranet environment (R&D portal) is seen as a key component for supporting and enabling information sharing and collaboration across the R&D organization. The current situations are that different R&D sites, which also include local functions, have different solutions to the issue of how to present information within the corporate intranet. The objective is to amplify productivity, support decision-making and organizational creativity by the improved management of scientific information in projects, across organizations (both internal and external). The company is planning to implement the new intranet structure in 2005.

This project has two interesting opportunities, First, it offers a new way of *exploiting* the information capital by creating a new IT infrastructure (i.e. creating a content management system) combined with a strategy of global standards (electronic formats for different data and documents), also in harmonization with regulatory demands, to overcome a fragmented structure of data and information sources within one framework. Thus, making information within the R&D organization easier to access combined with new search engines. Secondly, it opens up for *exploration* of information. The intranet framework (i.e. R&D portal) have inbuilt features such as *communities* combined with so-called *eRooms* (i.e. an electronic collaboration tool) to increase collaboration and informal sharing within teams and projects and different communities of practice. A community can be different groups of actors centred on a specific topic (e.g. technology, methodology aspects), or centred around skills, management groups, or on specific problem within, or outside, a project. Thus, the initiative can be serving two purposes. First, a top-down perspective on increasing effective access of information using *one* structured gateway to information. During the initial phases of the project, the target will be on handling primarily non-strategic information (e.g. organizational services, functional and skill descriptions, corporate news and so forth). Secondly, the project aims to increase the explorative capability of using strategic information (e.g. scientific and project-related information), which is essentially driven from a bottom up perspective.

Thus, the project offers interesting opportunities for exploiting (the traditional approach of using intranet) and exploring (e.g. operationalized in the concept communities) the company's entire information and intellectual capital. In addition, in order to be successful, the project has also identified a need to change cultural behaviours within the organization (e.g. such as information-sharing responsibilities and supporting informal networks).

Operational and short-term considerations

4 Supporting informal networking

Every organization carries out planned activities, and communication channels are necessary for these activities. But it is the unanticipated exchanges between employees who do not normally communicate with each other which often enable projects that have not been planned, to self-organize and move forward (Stacey, 1996). Organizational creativity involves flexibility and openness in the organization towards new ideas and attitudes. This flexibility in the organization is essential for promoting a kind of 'creative state of mind' to reduce conformism and mechanistic behaviour in the organization and to balance the high degree of projectification. These goals can be achieved by increased efforts to support and promote *informal networks*, which can balance effectiveness, standardization, and complex hierarchical organizational structures. Many pharmaceutical companies have developed a project organization model for new drug development activities that may serve as an impediment to such vital communication and interaction across organizational tiers and functions. One may argue that such organization leads to more cost-effective new drug development, while impeding organizational creativity and jeopardizing true innovation.

The creation of informal networks can be seen as a divergent process that balances the convergent process (i.e., projectification and reduced slack) and thus has the potential to create new arenas for dynamic, trans-disciplinary cooperation based primarily on intrinsic motivation. Moreover, actors in informal networks can also be seen as important *translators* in different interfaces between 'core creativity' and 'layer creativity' (see the model in Figure 3.2), thus enhancing the *connectivity* of ideas, experience, and knowledge among projects. This could mean, for example, that knowledge and the implications of a certain mechanism of the drug molecule (core creativity) could serve as a platform for ideas relevant for optimizing the formulation of the drug (e.g., different forms of administration) and new ways of performing clinical studies – not only for the particular project, but for other therapeutic areas, in particular. For example, co-workers who act as a translator about the discovery of new properties of the drug substance would be able to enhance the levels of meaning and connectivity between heterogeneous resources, skills, and projects related to finding new opportunities. The translator could get ideas rolling between projects and therapeutic areas. Moreover, in large and complex organizations informal networks can be seen as an alternative arena for communicating ideas.

This heterogeneity and interactivity, which symbolize the connectivity in new drug development process, has several practical implications. In order to promote organizational creativity to secure innovation in the pharmaceutical industry, financial and material support in research must be present, but is not sufficient. The shaping of new drugs requires networking heterogeneous elements both inside and outside the organization. Therefore, building communicative channels between researchers and other corporate members, and external actors such as academics, physicians, patients and regulators, are essential in mobilizing necessary elements or resources to achieve innovation. This interactive process is well illustrated by Hara (2003: 199):

> However if we incorporate the heterogeneity and networking of the process into the model, our pinball will become different from the familiar one: first there are many various balls (elements of drug technology) which are simultaneously bouncing of many various 'pins' (shapers of drug technology); second, the balls can combine together and split; third, the 'pins' are movable; forth, some pins are (human pins) have their intentions but others not; fifth, there is no strict distinction between 'balls' and 'pins'; sixth, there is no external player, and it is human pins that brings balls; seventh, some pins may exit the and others may enter. It could be also described as Football with various sticky balls, played between numerous teams at the same time in a rain forest.

On a practical level, how should one support informal networks in the pharmaceutical R&D? One way is to legitimize informal networking as an important skill, and to include it as one part of what employees are being evaluated on. However, one should be careful of putting extrinsic motivation factors in play for supporting informal networks. The very nature of informal networking is non-linearity, or what Stacey (1996) terms complex self-organizing adaptive systems, primarily driven by intrinsic motivation, so special attention is needed in creating these evaluation systems. However, feedback mechanisms for supporting informal networks can, on an individual level, be oriented on reflecting on learning and setting goals for future emerging networks and address how co-workers make contributions in networks. On a group level, one might create mentorships (which should also be replaced rather frequently) for informal networking, and encourage job rotation opportunities. Finally, on an organizational level, one could create training programmes specially designed to mentoring informal networks and/or build new routines for seminars in which co-workers from different skills, disciplines centred, for example, around new specific problems. These seminars may become new offshoots for emerging networks.

5 Co-worker training and safeguarding creative time – focusing on co-workers' scientific skills

The concept of time as related to *idea time* has been emphasized in different areas of creativity research (e.g., Ekvall and Ryhammar, 1999; Ekvall, 1997). Idea time is often related to incubation and necessary for the remote associations that tend to provide original ideas. One dimension in Ekvall's Creative Climate questionnaire, CCQ (e.g. Ekvall, 1996) is denoted 'idea time', which is described as the amount of time people can use (and do use) for elaborating new ideas and permits people to opportunities to explore. The reverse of this aspect is that every minute is booked and specified and the time pressure makes thinking outside the instructions and planned routines impossible (Isaksen *et al.*, 2001). Creative time can be used strategically, in 'let it happen' tactics, and it may be indicative of investments and intrinsic motivation (Runco, 1999). In this respect, idea time is ambiguous because it should not be taken for granted that everyone in an organization would be disturbed by deadlines, or working under pressure. An interesting comment from one senior researcher in AstraZeneca regarding idea time was:

> It's a strange thing; I think you have to be interested in what you are doing. I know it sounds strange, but these things don't go out of your mind. I have a different view from many colleagues. There are a lot of scientists that go around saying I don't have enough time to be creative. I need thinking time. If you're interested in the problem you can't stop thinking about it. You're unconsciously doing it. You're having a bath and you sit back and things start going around in your mind and out pop ideas. Where you need time isn't in the thinking, because thinking happens anyway. But where you need time is to test your hypotheses, test your ideas, either with what's known in the literature or with colleagues. (Senior Researcher, AstraZeneca)

Here, the important aspect of time is not securing time for thinking as such, but, rather, more time for testing and evaluating new ideas. However, what is at stake when the pharmaceutical industry has now become a mature large-scale industry in which profitability critically depends on launching new products in a timely fashion and efficiency and cost-effectiveness are in focus – for streamlining research and making it more productive (Pisano, 1997) – is time to test new ideas. One way forward is to integrate – and formalize – activities that can be legitimized as competence training in a specific skill, focused on a specific problem, but also to safeguard time for actually testing new ideas. A practical example would be to establish an ad hoc team or an informal network of a number of scientists that use the same kind of methodology (e.g. techniques like, chromatography, radioimmunoassay and so forth) to discuss problems and ideas how to solve certain issues in a limited time frame.

6 Revising performance measures in the firm

The predominant way of using performance measures in organizations is primarily centred on a controlling or control-based evaluation. This may be defined as a work evaluation characterized by the use of formalized standards and forms. Rules used to direct the individual to act in a certain way guide this assessment, which is less likely to involve sharing information and knowledge or exchanging ideas. Competence feedback, delivered in a controlling style, often makes external constraints salient. This implies that certain types of outcomes that the individual must obtain, or certain levels of creativity that he or she must achieve, are highlighted (Zhou, 1998). When confronted with competence feedback delivered in a controlling style, and interpreted by them as attempts at control, people generally experience feelings of external causality. They feel that there is someone else controlling their behaviour and actions. Thus it is likely that they interpret this style as attempts at inhibiting and restraining. This may increase extrinsic motivation at the expense of intrinsic motivation, and thus reduce creativity (Amabile, 1999b). Shalley and Perry-Smith (2001) found that there might be a connection between employees' self-rated creativity and how they are evaluated. Recent research suggests that situational factors can affect behaviour related to creativity in two ways: one *controlling* and the other *informational* (Shalley and Perry-Smith, 2001). Both have the potential to influence the way in which individuals perceive their own competence and self-determination for a specific task (e.g. Deci and Ryan, 1985; Ryan, 1982). The discussion concerning informational versus controlling evaluations furthermore resembles Zhou's (1998) notion of feedback style (informational versus controlling). The style of administrating rewards, rather than the rewards themselves, is the key issue in judging or perceiving rewards as either informational or controlling. Thus, in order to promote and evaluate efforts to enhance organizational creativity the usual method of evaluating co-workers is unsuitable.

An example of a performance measure that may have a potential to promote organizational creativity and be an indicator for intrinsic motivation, would be to introduce the concept of dialogue-base-evaluation which can be seen as a flexible, non-formalized evaluation of the work task (Sundgren *et al.*, 2005a). This concept involves making rewards more informational by acknowledging appropriate behaviour without using rewards to try to control behaviour. Dialogue-based evaluation is guided by and combined with giving information and thus creates an opening for the exchange of ideas and opinions and has the potential to uncover deeper meaning that necessitates exposing values, and, at least implicitly, keeping them under consideration.

One may argue that dialogue-based evaluation can bridge and reduce discrepancies between the *assumed* and *politically correct* culture versus the *enacted* and *true culture* and thus can be one way to manage creativity in an age of management control. On a practical level, this new performance measures faces several challenges when trying to put dialogue-based evaluation into practice: (i) it requires more time and effort than standardized methods; (ii) it requires a new kind of organizational capability that involves behavioural change on the part of individuals and the managerial system; and (iii) it challenges the traditional transactional leadership model in the sense that it emphasizes relations and requires a more open exchange of ideas rather than simply delivering according to fixed processes. Therefore, participants (managers and employees) must become more actively involved in providing information – thus creating an opening for exchange of ideas and opinions and establishing a dialogue that questions organizational values and norms. Revising performance measures in this direction could be a tool for retaining the intrinsic motivational focus, while simultaneously supporting the exploration of new ideas.

7 New thinking on leadership training

The systems perspective of creativity should not be restricted to research practices in new drug development; it should also include management practices to make sub-system interaction possible – to emphasize the need to create local techniques for translating research practices into managerial practices and reports. Control comes from the Latin term *contra rotulus*, which means 'against what is rolling'; one has to understand that if pharmaceutical research is to develop new ideas and solutions and get them rolling, then traditional management rationality, when interacting with such research, cannot be applied in the usual way.

Any new thinking in management training must deal with at least five factors. First, leaders need an improved ability to generate discussions and concrete action to support organizational creativity in the company.

This aspect involves a certain knowledge base about organizational creativity. But also being able to contribute to a richer picture of the current situation – using images and narratives of what organizational creativity is in new drug development. Secondly, being able to promote a balance between intrinsic and extrinsic motivation. Thirdly, develop skills in using political entrepreneurship and being able to cope with negative aspects of political behaviour. Fourthly, being able to not only understand the dynamics between stabilizers and destabilizers but also cope and act in such situations. For example, training programmes could be centred on how to understand and seek contractions, handle ambiguities and paradoxes, and develop the courage to see, and deal with, possibilities and threats. In dynamic terms, it is assumed that successful organizations are drawn to operate in the stable zone, were dynamics are determined to operate within stabilizers. Stacey (1996: 15) writes:

> The creative process that takes place at the edge of disintegration is inherently destructive, and paradoxical. It involves a cross-fertilization that can take place with mental symbols as well as with genetic material or digital code. True dialogue between human beings results in just such cross-fertilization, and we all know that true dialogue is usually uncomfortable, which is why we so rarely practice it. The creative process competition, which, as we know only to well, takes place in the medium of ideas, power, products . . . The creative process in human systems, therefore, is inevitable messy: it involves difference, conflict, fantasy, and emotion; it stirs up anger, envy, depression, and many other feelings. To remove the mess by inspiring us to follow some common vision, share the same culture, and pull together is to remove the mess that is the very raw material of creative activity.

Finally, a successful leadership able to promote organizational creativity must be based on a firm and broad understanding of the pharmaceutical research process. What would this leadership style look like in practice? One situation that might serve as a key illustration is taken from a real-life situation at one of the production facilities in AstraZeneca. The background was that the company expected a rapid increase in sales when one important product was granted OTC[44] approval in the US market in

---

[44] OTC, 'over the counter', are drug products, which do not need a prescription by a physician.

2004. As a consequence increased bulk production of the drug (active substance) became urgent. However, the outsourcing manufacturing plant of the bulk drug in the US could not handle this demand, and the problem was handed over to the production facility in Sweden. This demand came as rather a surprise to the Swedish production organization. The production plant director explains:

> We were demanded to increase our production capacity in a very time frame. I got a clear message from senior management to increase our production capacity with 280% [of tons substance per week]. When asking my process engineer of what would be feasible and what our maximum production capacity were. He said, 'We can increase up to 200%, but that's the theoretical limit'. When I told him, that we must increase our production capacity with 240% in two weeks, he said that I was mad. Before going public with the new target, I got an OK from senior management about the 240% target. You know within one week a number of different ad hoc teams had begun to evaluate different options, ideas and solutions in order to solve the problem. These cross-functional teams consisted of process engineers, medicinal chemists, production technicians, validation engineers and so forth. Within two weeks we had reach our target of new target of 240%. After two months we got the message from US operations that the initial target of 280% increased production per week was no longer necessary. Actually, I didn't think that we would have made it within such short time frame. But I deliberately set the new target higher of what my organization thought was doable. I actually increased the pace of finding new, and smarter, solutions on a complex problem, and I was convinced that we should have made it, but in a much longer time. My intention was to be humble about the task itself, but very goal oriented. I aimed to stretch my organization's creative capabilities, but not to a limit that would cause panic. You know, when we reached the target, there was a great sense of joy and proudness within the organization.

In this particular case, the manager was successful in carrying out the task. However, one might speculate that the underlying reason was because of an effective balance between destabilizers and stabilizers. Creative action was stimulated by deliberately destabilizing the organization by: (i) setting a higher goal for what the organization believed could be feasible; and (ii) introducing, and promoting, non-standardized ways of working (e.g. new informal networks, cross-functional teams). On the other hand, the manager stabilizes the organization with a clear goal, high impact for the business, and with a given time frame. By knowing the limits of what the organization could perform, and having the insight to push them a little harder, the manager destabilized the organization and promoted a kind of 'creative state of mind' to avoid conformism in the organization sufficiently to solve the problem. This example also demonstrates the importance of having courage, which, in this case, if he failed, may have had a negative impact on his career. Moreover, the manager also displayed political entrepreneurship in the sense that he created a commitment from senior management by setting a target that was sufficiently high to satisfy their demand.

There is actually a need for managers to cope with the anxiety of creative activity and being stuck in the zone of stabilizers. Stacey (1996: 281) writes, 'the antidote is to seek to keep the shadow system on the boil, to keep coming up with novel ways of doing this and then containing the anxiety that is raised'. From a senior management perspective, this type of leadership also needs to include the ability to bridge the gap between high-level strategic thinking, tactical operations, scientific excellence and dynamic, participative leadership. This means, for example, allowing the majority of key decisions to be aggregated from the organization rather than executive teams. Thus, new leadership training should be focused on to effectively create a dynamic equilibrium between stabilizers and stabilizers which share some characteristics that are in line what Collins (2001) terms level five leaders. Collins (2001) identifies five levels of leadership competency, from *Level One*, being a highly capable individual, to *Level Five*, being a leader who builds enduring success through a paradoxical blend of personal humility and professional will.

To summarize, there is no *best way* for management, but management and personal leadership should be guided to be more informal, flexible, inspirational, and visionary, but it also needs to encourage a highly challenging creative climate in order to facilitate a more efficient, decentralized decision making by skilled scientists form the guiding principle of operations, rather than traditional 'bottom-up lobbying' and executive decision-making as a control system.

Conclusion

An examination of the history of radical innovations in the pharmaceutical industry demonstrates the unpredictability of success and failure in organizations. One may argue that neither company size nor processes in the organization are the key issues when it comes to producing radical innovations (Sundgren, 2004). Organizational creativity, and in the best circumstances an innovation, in new drug development is about to create a dynamic equilibrium between destabilizers and stabilizers. It takes flexibility, intuition, connectivity, and courage to deal with and exploit unexpected findings along the path of research. And this demands a platform of trust and a certain amount of acceptance of deviance from standardized ways of doing things in the organization. One may argue that what is missing today in many large pharmaceutical companies is not new processes for effectiveness and efficiency, but *new thinking* on how to understand, promote, and manage organizational creativity.

Large pharmaceutical companies have several advantages compared to smaller firms: large resources for evaluating projects in the early phase, better opportunities for working across therapeutic areas, significantly more internal scientific data and information, and the economic power to go to market. But large pharmaceutical companies are also complex, rigid organizations. Cost-effectiveness and innovation are not necessarily opposing forces. At a time when the major companies have become equally good in mastering efficiency in new drug development, we argue that managing organizational creativity is going to create the *decisive competitive advantage* in the pharmaceutical industry. The ability of major players in the pharmaceutical industry to undertake radical innovations and contrive new, fruitful ways of organizing, depends upon the co-workers' opportunities to formulate and share new ideas within functions and project groups and between communities of practice, disciplines, skills, and functions in the company. Many pharmaceutical companies have developed a project organization model for new drug development activities that may serve as an impediment to such vital communication and interaction across organizational tiers and functions. One may propose that such organization leads to more cost-effective new drug development, while impeding organizational creativity and jeopardizing true innovation. Effective and safety is pivotal, but it is not enough.

In the case of the pharmaceutical industry, we argue that the dominant management approaches have been developed for the *logic of exploitation*, including organizational models, portfolio management tools, product strategies, project management systems, and the steering methods for minimizing deviations from these. A key challenge thus lies in more fully describing the *logic of exploration* in order to reveal ways of increasing the probability of organizational creativity and the likelihood of scientific product breakthroughs and the launch of new innovation lineages. This endeavour needs a pluralist approach to management, pluralist in the sense of Goodman's (1978: 4) use of the term:

> The pluralist, far from being anti-scientific, accepts the sciences at full value. His typical adversary is the monolithic materialist of phys-icalist who maintains that one system, physics, is preeminent and all-inclusive, such that every other version must eventually be reduced to it or rejected as false or meaningless.

The purpose of this chapter has been to serve as a platform consisting of different images and new concepts illustrating what *new thinking in management practice* is about in order to understand and promote management of organizational creativity – the *ex ante* process of innovation – and to avoid the formation of a predominantly *ex post* and instrumental rationality perspective. Finally, the arguments put forward in this book have centred on the pharmaceutical industry and new drug development, which demonstrates particular characteristics. These characteristics – an extremely regulated environment, high compliance with human and social ethics, a place at the forefront of advanced science and technology, high uncertainty in the research process, and being forced to produce radical innovations – serve as an interesting platform for further management and organizational research that will become relevant for other intensive innovation-based industries, such as automotive, aircraft, or telecom.

Summary and conclusions

This book has two aims: first, it seeks to criticize the existing organiza-tional creativity literature for presenting an unnecessarily simplified and, at times, even simplistic view of organizational creativity. This critique was presented in Part I of the book and evolved around the three themes of theory, epistemology and methodology. While some literature on organizational creativity is formulated on the basis of clearly positioned theoretical assumptions and objectives, some literature is more fuzzy and almost distinguishable from journalistic work. When this literature presents too rosy a view of what organizational creativity can enable and what problems the construct of organizational creativity can handle, we have spoken of it in terms of kitsch – that is, a certain aesthetic, ethic, and political position assuming anthropocentrism and largely adhering to consensus theories of society. The first part of the book generally encourages a more self-reflexive and self-critical view of creativity wherein organizational creativity is not only advocated and called into question as a universal remedy for a number of managerial and organizational evils such as bureaucratization, loss of innovative capacities, and the inability to exploit and explore intellectual capital, but is also where organizational creativity is examined as a set of inter-related and co-dependent resources and practices that jointly constitute the organization's creative capabilities. In much of the organizational creativity literature, creativity remains very much a *black box* devoid of social and human content – that is, a series of practices and activities comprising and drawing on politics, communication, emotionality, disappointments, negotiations, and so forth. To date, research on organizational creativity remains a subset within the large and hetero-geneous organization theory and management literature, and the research on organizational creativity is mostly published in specialized journals. In order to promote research on organizational creativity, a more self-critical attitude may be helpful.

The second aim of the book is to point to a number of practical issues pertaining to the day-to-day management of organizational creativity that we believe has been, if not marginalized, at least not sufficiently theorized and empirically investigated in the organizational creativity literature. The use and function of technology in organizational creativity work, leadership practices, and complementary rationalities such as intuition are the three facets of organizational creativity discussed in Part II of the book. Technology is in many cases and in many industries and companies an indispensable resource in work that draws on organi-zational creativity. The merits of technology are numerous and its influence on society is pervasive and hard to underrate. Yet organizational creativity is by no means determined or inextricably entangled with superior use of technology per se. Technology is rather constituting the framework wherein creative ideas and thought can be formulated. Failing to disentangle organizational creativity and technology easily leads to the belief that it is *technology* per se that enable new thinking and new ideas. Therefore, technology needs to be examined carefully prior to large-scale investments; social, cultural, and emotional conse-quences of technology deserve detailed consideration.

Management theory and practice tend to celebrate rationalities that are rooted in transparent and easily understood regimes of representation; a close connection between saying and doing is praised in management thinking. But this strong emphasis on language and representation as a medium capable of mirroring complex condition and processes is becoming problematic. For instance, in knowledge-intensive companies

(Robertson and Swan, 2003; Starbuck, 1992), accounting practices are not longer capable of adequately representing the intellectual capital of the firm (Mouritsen, Larsen and Bukh, 2001). The inability to capture underlying resources and assets is then constituting a major challenge for managers and actors in the financial markets because one can no longer be assured what the true organizations book value may be. In a similar manner, organizational creativity is only partially captured by the pre-existing and shared vocabularies in a company. In our account, thinking that defies the day-to-day vocabulary is referred to as *intuition*. While intuition is not capable of establishing itself as a legitimate resource in organizations because of its place outside of language, it still plays a central role in organizational creativity. As a consequence, a broader outlook on how organizational creativity is unfolding in the course of action in practices and activities may enable a recognition of rationalities such as intuition. Finally, leadership is another major concern for organizations and companies dependent on their ability to exploit their capacities for organizational creativity. The issue of leadership in creative settings implies two risks: Leadership may either be overrated as what is the single most important precursor for organizational creativity, or, in the diametrically opposed position, leadership may be degraded as what is not only irrelevant for creative work but also what is directly detrimental for creativity. The former idea draws on what is at times called *managerialism* (Kitchener, 2002; Young, 1999; Deetz, 1992; Thompson, 1969), the ideology portraying systematic management and organization as a universal strategy capable of dealing with all possible social problems and challenges.

In managerialist thinking, leadership is a sacred activity effectively sorting things out. The latter view of leadership is rooted in what Ford (1995a) calls the romance of creativity wherein 'great men and women' (mostly men, unfortunately) are operating in isolation from all external forces and expectations. Neither of these two end-positions are adequate descriptions of how leadership influence and support organizational creativity. In our view, leadership matter a great deal but it cannot substitute for the creative work per se. Instead, leadership is what is supporting and reinforcing organizational creativity and therefore it plays an important role in organizations.

In this final chapter, we have aimed to make the point that these different discussions and perspectives on organizational creativity are not academic and sophistic hair-splitting activities; rather, they have practical implications for organizations. However, it is complicated to formulate universally applicable rules and suggestions about something – in our case organizational creativity – that is, of necessity, local, situational, contingent, and context-bound. Organizational creativity is what is performed in specific communities of practice, operating on jointly formulated research objectives and under the influence of various local financial, temporal, and resource-based constraints. Advice and suggestions are therefore incapable of embodying all such contingencies, and consequently rules and suggestions are of necessity expressed in very general terms. Still, it is our intention not only to offer a series of critical accounts, but also to make a contribution, perhaps small, insignificant, or ephemeral, but nevertheless a contribution to the understanding of how organizational creativity can influence and develop organizations.

Bibliography

Abbey, A. and Dickson, J. (1983). R&D work climate and innovation in semiconductors. *Academy of Management Journal*, 26: 362–8.

Abrahamson, E. (1991). Managerial fads and fashions: the diffusion and rejection of innovations. *Academy of Management Review*, 16: 586–612.

Achilladelis, B. (1999). Innovation in the pharmaceutical industry. In R. Landau, B. Achilladelis and A. Scriabine (eds), *Pharmaceutical Innovation*. Philadelphia: Chemical Heritage Press, pp. 1–147.

Agor, W. (1989). *Intuition in Organizations: Leading and Managing Productively*. Newbury Park, CA: Sage.

Ahuja, G. and Lampert, C.M. (2001). Entrepreneurship in the large corporation: a longitudinal study of how established firms create breakthrough inventions. *Strategic Management Journal*, 22: 521–43.

Albert, R.S. and Runco, M.A. (1999). A history of research on creativity. In R.J. Sternberg (ed), *Handbook of Creativity*. Edinburgh: Cambridge University Press, pp. 16–31.

Allison, G.T. (1971). *Essence of Decision: Examining the Cuban Missile Crisis*. New York: HarperCollins.

Alvesson, M. (1992). Leadership as social integrative action: a study of a computer consultancy company. *Organizational Studies*, 13 (2): 185–210.

Alvesson, M. (1996). *Communication, Power, and Organization*. Berlin: De Gruyter.

Alvesson, M. (2001). Knowledge work: ambiguity, image and identity. *Human Relations*, 54 (7): 863–86.

Alvesson, M. and Svenningsson, S. (2003). Good visions, bad micro-management and ugly ambiguity: contradiction of (non)-leadership in a knowledge intensive organization. *Organization Studies*, 24 (6): 961–88.

Amabile, T.M. (1982). Social psychology of creativity: a consensual assessment technique. *Journal of Personality and Social Psychology*, 43: 997–1013.

Amabile, T.M. (1988). A model of creativity and innovation in organizations. *Research in Organizational Behavior*, 10: 123–67.

Amabile, T.M. (1996). *Creativity in Context: Update to the Social Psychology of Creativity*. Boulder, CO: Westview.

Amabile, T.M. (1997). Motivating creativity in organizations: on doing what you love and loving what you do. *California Management Review*, 40 (1): 39–58.

Amabile, T.M. (1999a). How to kill creativity. *Harvard Business Review*, September–October, 77–87.

Amabile, T.M. (1999b). *Creativity in Context: Update to the Social Psychology of Creativity*. Boulder, CO: Westview.

Amabile, T.M. and Conti, R. (1999). Changes in the work environment for creativity during downsizing. *Academy of Management Journal*, 42 (6): 630–40.

Amabile, T.M., Conti, R., Coon, H., Lasenby, J. and Herron, M. (1996). Assessing the work environment for creativity, *Academy of Management Journal*, 39 (5): 1154–1184.

Amabile, T.M. and Gryskiewicz, N.D. (1989). The creative environment scales: work environment inventory. *Creativity Research Journal*, 2: 231–53.

Amabile, T.M., Schatzel, E.A., Moneta, G.B. and Kramer, S.J. (2004). Leader behaviors and the work environment for creativity: Perceived leader support. *The Leadership Quarterly*, 15: 5–32.

Ambrosini, V. and Bowman, C. (2002). Mapping successful organizational routines. In Anne S. Huff and M. Jenkins (eds), *Mapping Strategic Knowledge*. London, Thousand Oaks and New Delhi: Sage.

Andriopoulos, C. (2001). Determinants of organizational creativity: a literature review. *Management Decision*, 39 (10): 834–40.

Andriopoulos, C. (2003). Six paradoxes in managing creativity: an embracing act. *Long Range Planning*, 36: 375–88.

Angel, M. (2004). *The Truth about the Drug Companies*. New York: Random House.

Ansell Pearson, K. (1999). *Geminal Life: The Difference and Repetition of Deleuze*, London: Routledge.

Ansell Pearson, K. (2002). *Philosophy and the Adventures of the Virtual: Bergson and the Time of Life*, London and New York: Routledge.

Appelbaum, S.H. and Gellagher, J. (2000). The competitive advantage of organizational learning. *Journal of Workplace Learning*, 12 (2): 20–56.

Argyris, C. and Schön, D. (1993). *Knowledge for Action*. San Francisco: Jossey-Bass.

Aristotle (1986). *De Anima*. London: Penguin.

Armitage, J. (2001). *Virilio Live: Selected Interviews*. London, Thousand Oaks and New Delhi: Sage.

Astley, W.G. and Sachdeva, P.S. (1984). Structural sources of intraorganizational power: a theoretical synthesis. *Academy of Management Review*, 9 (1): 104–13.

Atkinson, P. and Coffrey, A. (2003). Revisiting the relationship between particpant observations and interviewing. In Jaber F. Gubrium and James A. Holstein (eds), *Postmodern Interviewing*. London, Thousand Oaks and New Delhi: Sage, pp. 109–22.

Augier, M. and Thanning Vendelø, M. (1999). Networks, cognitions and management of tacit knowledge. *Journal of Knowledge Management*, 3 (4): 252–61.

Bachelard, G. (1964). *The Poetics of Space*, trans. Maria Jolas, Boston: Beacon Press.

Bachelard, G. (1984). *The New Scientific Spirit*, Boston: Beacon Press.

Badiou, A. (1999). *Deleuze: The Clamour of Being*, Minneapolis: University of Minnesota Press.

Bain, A. (1998). Social defenses against organizational learning. *Human Relations*, 51 (3): 413–29.

Ball, K. and Carter, C. (2002). The charismatic gaze: everyday leadership practices of the 'new' manager. *Management Decision*, 40 (6): 552–65.

Ball, K. and Wilson, D.C. (2000). Power, control and computer-based performance monitoring: repertoires, resistances and subjectivities. *Organization Studies*, 21 (3): 539–65.

Barley, S.R. (1986). Technology as an occasion of structuring: evidence from observations of CT scanners and the social order of radiology departments. *Administrative Science Quarterly*, 31: 78–108.

Barley, S.R. (1990). The alignment of technology and structure through roles and networks. *Administrative Science Quarterly*, 35: 61–103.

Barley, S.R. and Kunda, G. (1992). Design and devotion: surges of rational and normative ideologies of control in managerial discourse. *Administrative Science Quarterly*, 37: 363–99.

Barney, J. (1991). Firm resources and sustained competitive advantage. *Journal of Management*, 17: 99–120.

Barthes, R. (1977). *Image, Music, Text*. London: Fontana.

Bass, B.M. (1990). *Bass & Stogdils's Handbook of Leadership*, 3rd edition. New York: The Free Press.

Bataille, G. (1983). *Visions of Excess*. Manchester: Manchester University Press.

Bauman, Z. (1999). *Culture as Praxis*. London, Thousand Oaks and New Delhi: Sage.

Baumard, P. (1999). *Tacit Knowledge in Organizations*. London: Sage.

Bay, T. (1998). *AND . . . AND . . . AND: Reiterating Financial Derivation*, School of Business, Stockholm University.

Bechara, A., Damasio, H., Tranel, D. and Damasio, A.R. (1997). Deciding advantageously before knowing the advantageous strategy. *Science*, 275: 1293–1295.

Bell, D. (1973). *The Coming Post-Industrial Society*. New York: Basic Books.

Benhabib, S. (2002). *The Claims of Culture: Equality and Diversity in the Global Era*. Princeton and Oxford: Priceton University Press.

Benjamin, W. (1999). *The Arcades Project*, trans. H. Eiland and K. McLaughlin, Cambridge: The Belknap Press.

Bentham, J. (1995). *The Panopticon Writings*. London: Verso.

Bergson, H. (1919/1992). *The Creative Mind: An Introduction to Metaphysics*. New York: Citadel Press.

Bergson, H. (1911/1998). *Creative Evolution*. Mineola: Dover Publishers.

Bergson, H. (1912/1999). *An Introduction to Metaphysics*. Indianapolis: Hackett.

Bertalanffy, L. (1968). *General System Theory: Foundations, Development, Applications*. New York: George Braziller.

Best, S. and Kellner, D. (1991). *Postmodern Theory: Critical Interrogations*. London: Macmillan.

Bijker, W.E. (1995). *Of Bicycles, Bakelites, and Bulbs: Toward a Theory of Sociotechnical Change*. Cambridge and London: The MIT Press.

Björkman, H. and Sundgren, M. (2005). Political entrepreneurship in action research: learning from two cases. *Journal of Organization Learning and Change*.

Blackler, F., Crump, N. and McDonald, S. (1999). Managing experts and competing through innovation: an activity theoretical analysis. *Organization*, 6 (1): 5–31.

Boden, D. (1994). *The Business of Talk: Organizations in Action*. Cambridge: Polity Press.

Boden, M. (1996). *Dimensions of Creativity*. London: Bradford Books.

Bohm, D. (1998). *On Creativity*, ed. Lee Nichol. London: Routledge.

Bohm, D. and Peat, D.F. (1989). *Science, Order and Creativity*. London: Routledge.

Boisot, M.H. (1998). *Knowledge Assets: Securing Competitive Advantage in the Information Economy*. Oxford: Oxford University Press.

Boje, D.M. (1991). The storytelling organization: a study of story performance in an office supply firm. *Administrative Science Quarterly*, 36: 106–26.

Boje, D.M. (2001). *Narrative Methods for Organization Research & Communication Research*. London, Thousand Oaks and New Delhi: Sage.

Bolter, J.P. (1991). *Writing Space: The Computer, Hypertext, and the History of Writing*. Hillsdale, Hove and London: Lawrence Erlbaum.

Boudreau, M.-C., Loch, K.D., Robey, D. and Straud, D. (1998). Going global: using information technology to advance the competitiveness of the virtual transnational organization. *Academy of Management Executive*, 12 (4): 120–8.

Bougen, P.D. and Young, J.J. (2000). Organizing and regulating as rhizomatic lines: bank fraud and auditing. *Organization*, 7 (3): 403–26.

Bourdieu, P. (1996). *On Television*. New York: The New Press.

Bourdieu, P. and Haake, H. (1995). *Free Exchange*. Cambridge: Polity Press.

Bourdieu, P. and Passeron, J.-C. (1977). *Reproduction in Education, Society, and Culture*. London: Sage.

Bowers, K.S. (1973). Situationism in psychology: an analysis and a critique. *Psychological Review*, 80: 307–36.

Bowker, G. (1995). Manufacturing truth: the development of industrial research. In M. Serres (ed.), *A History of Scientific Thought*. Oxford: Blackwell.

Bowker, G.C. and Star, S.L. (1999). *Sorting Things Out: Classification and Its Consequences*. Cambridge and London: The MIT Press.

Boyce, M.E. (1995). Collective centrings and collective sense-making in the stories and storytelling of one organization. *Organization Studies*, 16 (1): 107–37.

Braidotti, R. (1994). *Nomadic Subjects: Embodiment and Sexual Difference in Contemporary Feminist Theory*. New York: Columbia University Press.

Braidotti, R. (1997). Meta[l]morphoses. *Theory, Culture & Society*, 14 (2): 67–80.

Bras, D.J. (1995). Creativity: it's all in your social network. In C.M. Ford and D.A. Gioia (eds), *Creative Action in Organizations: Ivory Tower Visions and Real World Voices*. Thousand Oaks, CA: Sage Publications, pp. 94–9.

Brewis, J. and Linstead, S. (2000). *Sex, Work and Sex Work*. London and New York: Routledge.

Broch, H. (1933/1968). Notes on the problem of kitsch. In G. Dorfles (ed.), *Kitsch: An Anthology of Bad Taste*. London: Studio Vista, pp. 49–76.

Brown, S.D. and Lightfoot, G. (2002). Presence, absence, and accountability: e-mail and the mediation of organizational memory. In S. Woolgar (ed.), *Virtual Society? Technology, Cyberbole, Reality*. Oxford and New York: Oxford University Press.

Bruner, J. (1986). *Actual Minds, Possible Worlds*, Cambridge: Harvard University Press.

Bruner, J. (1990). *Acts of Meaning*. Cambridge: Harvard University Press.

Bryans, P. and Smith, R. (2000). Beyond training: reconceptualizing learning at work. *Journal of Workplace Learning*, 12 (6): 228–35.

Bryman, A. (1996). Leadership in organizations. In S.R. Clegg, C. Hardy and W.R. Nord (eds), *Handbook of Organization Studies*. London: Sage.

Bryman, A. (2000). Telling technological tales. *Organization*, 7 (3): 455–75.

Buchanan, D. and Badham, R. (1999). *Power, Politics, and Organizational Change: Winning the Turf Game*. London: Sage.

Buchanan, I. (1997). Deleuze and cultural studies. *The South Atlantic Quarterly*, 96 (3): 483–97. Special Issue: *A Deleuzian Century?*

Bunce, D. and West, M.A. (1996). Stress management and innovations at work. *Human Relations*, 49 (2): 209–32.

Burke, T.E. (2000). *Whitehead's Philosophy*. London: Greenwich Exchange.

Burns, T. and Stalker, G.M. (1961). *The Management of Innovation*. London: Tavistock Publications.

Burrell, G. and Morgan, M. (1979). *Sociological Paradigms and Organizational Analysis*. London: Heinemann.

Burton-Jones, A. (1999). *Knowledge Capitalism: Business, Work, and Learning in the New Economy*. Oxford: Oxford University Press.

Calás, M. and Smircich, L. (1991). Voicing seduction to silence leadership. *Organization Studies*, 12 (4): 567–602.

Capral, F. (1996). *The Web of Life*. New York: Anchor Books/Doubleday.

Cardinal, L.B. (2001). Technological innovation in the pharmaceutical industry: the use of organizational control in managing research and development. *Organization Science*, 12 (1): 19–36.

Carlsson, S. (1951). *Executive Behaviour: A Study of the Work Load and the Working Methods of Managing Directors*. Stockholm: Strömbergs.

Carr, A. (2001). Understanding emotion and emotionality in a process of change. *Journal of Organization Change Management*, 14 (5): 421–34.

Castoriadis, C. (1987). *The Imaginary Institutions of Society*, trans. K. Blamey. Cambridge: Polity Press.

Cheng, Y.-T. and van de Ven, A.H. (1996). Learning the innovation journey: order out of chaos? *Organization Science*, 7 (6): 593–605.

Chia, R. (1995). From modern to postmodern organizational analysis. *Organization Studies*, 16 (4): 579–604.

Chia, R. (1998). From complexity of science to complex thinking: organization as simple location. *Organizations*, 5 (3): 341–469.

Chia, R. (1999). A 'rhizomatic' model of organizational change and transformation: perspective from a metaphysics of change. *British Journal of Management*, 10: 209–27.

Chia, R. (2003). From knowledge-creation to the perfecting of action: Tao, Basho and pure experience as the ultimate ground of knowing. *Human Relations*, 56 (8): 953–81.

Christensen, C. and Raynor, M. (2003). *The Innovator's Solution*. Cambridge, MA: Harvard Business School Press.

Clifford, J. (1997). *Routes: Travel and Translation in the Late Twentieth Century*. Cambridge: Harvard University Press.

Coghlan, D. and Brannick, T. (2001). *Doing Action Research in Your Own Organization*. London: Sage.

Cohen, W.M. and Levinthal, D.A. (1990). Absorptive capacity: a new perspective on learning and innovation. *Administrative Science Quarterly*, 35: 128–52.

Collins, J. (2001). *Good to Great*. New York: Harper Collins.

Contu, A., Grey, C. and Örtenblad, A. (2003). Against learning. *Human Relations*, 56 (8): 931–52.

Contu, A. and Willmott, H. (2003). Re-embedding situatedness: the importance of power relations in learning theory. *Organization Science*, 14 (3): 283–96.

Cook, S.D. and Yanow, D. (1993). Culture and organizational learning. *Journal of Management Inquiry*, 2 (4): 373–90.

Cooper, L.P. (2003). A research agenda to reduce risk in new product development through knowledge management: a practitioner perspective. *Journal of Technology Management*, 20: 117–40.

Cropley, A.J. (1999). Definitions of creativity. In M.A. Runco and S.R. Pritzker (eds), *Encyclopaedia of Creativity*. London: Academic Press, pp. 511–24.

Csikszentmihalyi, M. (1988). Society, culture, and person: a systems view of creativity. In R.J. Sternberg (ed.), *The Nature of Creativity: Contemporary Psychological Perspectives*. New York: Cambridge University Press, pp. 325–39.

Csikszentmihalyi, M. (1994). The domain of creativity. In D.H. Feldman, M. Csikszentmihalyi and H. Gardner, *Changing the World: A Framework of the Study of Creativity*. Westport: Praeger, pp. 135–58.

Csikszentmihalyi, M. (1996). *Creativity: Flow and the Psychology of Discovery and Invention*. New York: Harper Collins.

Csikzentmilhayi, M. (1999). Implications of a systems perspective. In R.J. Sternberg (ed.), *Handbook of Creativity*. Cambridge: Cambridge University Press, pp. 313–28.

Csikszentmihalyi, M. and Sawyer, K. (1995). Shifting the focus from individual to organizational creativity. In C.M. Ford and D.A. Gioia (eds), *Creative Action in Organizations: Ivory Tower Visions and Real World Voices*. Thousand Oaks, CA: Sage Publications, pp. 167–72.

Cunliffe, A.L., Luhman, J.T. and Boje, D.M. (2004). Narrative temporality: implications for organizational research. *Organization Studies*, 25 (2): 261–86.

Currie, G. and Brown, A.D. (2003). A narratological approach to understanding processes of organizing in a UK hospital. *Human Relations*, 56 (5): 563–86.

Czarniawska, B. (1998). *A Narrative Approach to Organization Studies*. Thousand Oaks, London and New Delhi: Sage.

Czarniawska, B. (2004). *Narratives in Social Science Research*. London, Thousand Oaks and New Delhi: Sage.

Czarniwska, B. and Mazza, C. (2003). Consulting as a liminal space. *Human Relations*, 56 (3): 267–90.

Daft, R.L. and Weick, K.E. (1984). Toward a model of organizations as interpretation systems. *Academy of Management Review*, 9 (2): 284–95.

Daly, H.E. and Cobb, J.B. (1990). *For the Common Good: Redirecting the Economy Towards Community, the Environment and a Sustainable Future*. Boston: Beacon.

Damanpour, F. (1998). Innovation type, radicalness, and the adoption process. *Communication Research*, 15: 545–67.

D'Aveni, R.A. (1994). *Hypercompetition*. New York: Free Press.

Davenport, T.H. and Prusak, L. (1998). *Working Knowledge: How Organizations Manage What They Know*. Boston: Harvard Business School Press.

De Bono, E. (1985). *Six Thinking Hats*. Boston: Little Brown.

De Bono, E. (1992). *Serious Creativity: Using the Power of Lateral Thinking to Create New Ideas*. New York: Harper Collins.

Deci, E.L. and Ryan, R.M. (1985). *Intrinsic Motivation and Self-determination in Human Behavior*. New York: Plenum.

Deetz, S.A. (1992). *Democracy in an Age of Corporate Colonialization*. Albany: State of New York University Press.

Deleuze, G. (1988a). *Spinoza: Practical Philosophy*. San Francisco: City Lights Books.

Deleuze, G. (1988b). *Bergsonism*. New York: Zone Books.

Deleuze, G. (1990a). *Expressionism in Philosophy: Spinoza*. New York: Zone Books.

Deleuze, G. (1990b). *Negotiations*. New York: Columbia University Press.

Deleuze, G. (1993). *The Fold: Leibniz and the Baroque*. Minneapolis: University of Minnesota Press.

Deleuze, G. (1997). *Essays Critical and Clinical*. Minneapolis: University of Minnesota Press.

Deleuze, G. and Guattari, F. (1988). *A Thousand Plateaus: Capitalism and Schizophrenia*. Minneapolis: University of Minnesota Press.

Deleuze, G. and Guattari, F. (1995). *What is Philosophy?* London, Verso.

De Peuter, J. (1998). The dialogics of narrative identity. In M.M. Bell and M. Gardiner (eds), *Bakhtin and the Human Sciences*. London, Thousand Oaks and New Delhi: Sage.

Dill, D.D. and Pearson, A.W. (1984). The effectiveness of project managers: Implications of political model. *IEEE Transactions on Engineering Management*, 31 (3): 138–46.

Dimenäs, E., Glise, H. and Simon, K. (2000). Strategic aspects of clinical R&D in a continuously changing environment. *Proceedings of IRIS 23*, Laboratorium for Interaction Technology, 2000.

Dixon, N.M. (2000). *Common Knowledge*. Cambridge: Harvard Business School Press.

Dodgson, M. (1993). Organizational learning: a review of some of the literature. *Organization Studies*, 14 (3): 373–94.

Dohlsten, M. (2003). Executive commentary on Part I. In N. Adler, A.B Shani and A. Styhre (eds), *Collaborative Research in Organizations*. Thousand Oaks CA: Sage, pp. 75–8.

Donnellon, A. (1996). *Team Talk: Listening Between the Lines to Improve Team Performance*. Boston: Harvard Business School Press.

Doolin, B. (2002). Enterprising discourse, professional identity and the organizational control of hospital clinicians. *Organization Studies*, 23 (3): 369–90.

Dorabjee, S., Lumley, C.E. and Cartwright, S. (1998). Culture, innovation and successful development of new medicines – an exploratory study of the pharmaceutical industry. *Leadership & Organization Development Journal*, 19 (4): 199–210.

Dorfles, G. (1968). *Kitsch: An Anthology of Bad Taste*. London: Studio Vista.

Dougherty, D. (1999). Organizing for innovation. In S.R. Clegg, C. Hardy and W.R. Nord (eds), *Managing Organizations*. London: Sage.

Dougherty, D. and Hardy, C. (1996). Sustained product innovation in large mature organizations: Overcoming innovation-to-organization problems. *Academy of Management Journal*, 39 (5): 1120–1153.

Drazin, R. (1990). Professionals and innovation: structural-functional versus radical structural perspectives. A multilevel perspective. *Journal of Management Studies*, 27: 245–63.

Drazin, R., Glynn, M.A. and Kazanjian, R.K. (1999). Multilevel theorizing about creativity in organizations: A sense-making perspective. *Academy of Management Review*, 24: 286–307.

Dreher, G.F. (2003). Breaking the glass-ceiling: the effects of sex ratios and work-life programs on female leadership at the top. *Human Relations*, 56 (5): 541–62.

Drews, J. (1997). Strategic choices facing the pharmaceutical industry: A case for innovation. *Drug Discovery Today*, 2 (2): 72–8.

Drews, J. (1998). Innovation deficit revisited: reflections on the productivity of pharmaceutical R&D. *Drug Discovery Today*, 3 (11): 491–4.

Drews, J. (2003). Strategic trends in the drug industry. *Drug Discovery Today*, 8 (9): 411–20.

Driver, M. (2002). The learning organization: Foucaultian gloom or Utopian sunshine. *Human Relations*, 55 (1): 33–53.

Drucker, P.F. (1955). *The Practice of Management*. Melbourne, London and Toronto: Heinemann.

Dupré, J. (1993). *The Disorder of Things: Metaphyical Foundations of the Disunity of Science*. Cambridge and London: Harvard University Press.

Durkheim, E. (1895/1938). *The Rules of Sociological Method*. Glencoe, IL: Free Press.

Easterby-Smith, M. (1997). Disciplines of organizational learning: contributions and critiques. *Human Relations*, 50 (8): 1085–1113.

Easton, D. (1965). *A Framework for Political Analysis*. Englewood Cliffs, NJ: Prentice Hall.

Eco, U. (1989). The structure of bad taste. In *The Open Work*, trans. A. Cancogni. London: Hutchinson Radius.

Edenius, M. and Hasselbladh, H. (2002). The balanced scorecard as an intellectual technology. *Organization*, 9 (2): 249–73.

Edmonson, A.C., Bohmer, R.M. and Pisano, G.P. (2001). Disrupted routines: team learning and new technology implementation in hospitals. *Administrative Science Quarterly*, 46: 685–716.

Ehrenreich, B. (1989). *Fear of Falling: The Inner Life of the Middle Class*. New York: Harper Perennial.

Ekvall, G. (1987). The climate metaphor in organizational theory. In I.B.M. Bass and P.J.D. Drent (eds), *Advances in Organizational Psychology: An International Review*. Newbury Park, CA: Sage Publications, pp. 177–90.

Ekvall, G. (1995). The creative climate: its determinants and effects at a Swedish University. *Creativity Research Journal*, 12 (4): 303–10.

Ekvall, G. (1996). Organizational climate for creativity and innovation. *European Journal of Work and Organizational Psychology*, 5 (1): 105–23.

Ekvall, G. (1997). Organizational conditions and levels of creativity. *Creativity and Innovation Management*, 6 (4): 195–205.

Ekvall, G. (1999). *Creative Climate*. In S.R. Pritzker (ed.), *Encyclopaedia of Creativity*. London: Academic Press, pp. 403–12.

Ekvall, G. and Ryhammar, L. (1999). The creative climate: its determinants and effects at a Swedish university. *Creativity Research Journal*, 12 (4): 303–10.

Ellul, J. (1964). *The Technological Society*. New York: Vintage.

Eysenck, H.J. (1996). The measurement of creativity. In M.A. Boden (ed.), *Dimensions of Creativity*. London: Bradford Books.

Ezzell, C. (2002). Proteins rule: biotech's latest mantra is "proteomics," as it focuses on how dynamic networks of human proteins control cells and tissues. *Scientific American*, April: 40–5.

Feist, G.J. and Runco, M.A. (1993). Trends in the creativity research: an analysis of research in the Journal of Creative Behaviour (1967–1989). *Creativity Research Journal*, (6): 271–86.

Feldman, D.H. (1999). The development of creativity. In R.J. Sternberg, (ed.), *Handbook of Creativity*. Cambridge, UK: Cambridge University Press, pp. 169–88.

Feldman, D.H., Csikszentmihalyi, M. and Gardner, H. (1994). *Changing the World: A Framework of the Study of Creativity*. Westport: Praeger, pp. 135–58.

Feldman, S.P. (2004). The culture of objectivity: quanitification, uncertainty, and the evaluation of risk at NASA. *Human Relations*, 57 (6): 691–718.

Festinger, L. (1957). *A Theory of Cognitive Dissonance*. Stanford, CA: Stanford University Press.

Feyerabend, P. (1999). *Conquest of Abundance: A Tale of Abstraction versus Richness of Being*. Chicago and London: The University of Chicago Press.

Fineman, S. (ed.) (1993). *Emotions in Organizations*. London: Sage.

Fiol, C.M. and Lyles, M.A. (1985). Organizational learning. *Academy of Management Review*, 10 (4): 803–13.

Fleck, L. (1979). *Genesis and Development of a Scientific Fact*. Chicago and London: Chicago University Press.

Flynn, J. and Straw, M. (2003). Lend me your Wallet: the effect of charismatic leadership on external support or an organization. *Strategic Management Journal*, 25: 309–30.

Ford, C.M. (1995a). Creativity is a mystery. In C.M. Ford and D.A. Gioia (eds), *Creative Action in Organizations: Ivory Tower Visions and Real World Voices*. Thousand Oaks: Sage Publications, pp. 12–53.

Ford, C.M. (1995b). A multi-domain model of creative action taking. In C.M. Ford and D.A. Gioia (eds), *Creative Action in Organizations: Ivory Tower Visions and Real World Voices*. Thousand Oaks: Sage Publications, pp. 330–55.

Ford, C.M. (1996). A theory of individual creative action in multiple social domains. *Academy of Management Review*, 21: 1112–1142.

Ford, C.M. (2002). The futurity of decisions as a facilitator of organizational creativity and change. *Journal of Organizational Change Management*, 15 (6): 635–46.

Ford, C.M. and Gioia, D.A. (1995). Multiple visions and multiple voices: academics' and practitioners' conceptions of creativity in organizations. In C.M. Ford and D.A. Gioia (eds), *Creative Action in Organizations: Ivory Tower Visions and Real World Voices*. Thousand Oaks: Sage Publications, pp. 3–11.

Ford, C.M. and Gioia, D.A. (2000). Factors influencing creativity in the domain of managerial decision making. *Journal of Management*, 26 (4): 705–32.

Foss, N.J. (1996). Knowledge-based approaches to the theory of the firm: some critical remarks. *Organization Science*, 7 (5): 470–6.

Foucault, M. (1980). *Power/Knowledge*. New York: Harvester Wheatsheaf.

Fox Keller, E. (2002). *Making Sense of Life: Explaining Biological Development with Models, Metaphors, and Machines*. Cambridge and London: Harvard University Press.

Frank, T. (1997). *The Conquest of Cool: Business Culture, Counterculture and the Rise of Hip Consumerism*. Chicago and London: The University of Chicago Press.

Friedman, L.M., Furberg, C.D. and DeMets, D.L. (1985). *Fundamentals of Clinical Trials*, 2nd edn. St Louis: Mosby Year Book.

Friedman, V.J., Lipshitz, R. and Overmeer, W. (2001). Creating conditions for organizational learning. In M. Dierkes, A. Berthon, J. Child and I. Nonaka (eds), *Handbook of Organizational Learning & Knowledge*. Oxford: Oxford University Press.

Frost, P.J. and Egri, C.P. (1991). The political process of innovation. *Research in Organizational Behavior*, 13: 229–95.

Gabriel, Y. (2000). *Storytelling in Organizations: Facts, Fictions, and Fantasies*. Oxford: Oxford University Press.

Gabriel, Y. (2004). Introduction. In G. Yannis (ed.), *Myths, Stories, and Organizations: Premodern Narratives of Our Time*, Oxford and New York: Oxford University Press, pp. 1–9.

Galunic, C.D. and Rodan, S. (1998). Resource recombinations in the firm: knowledge structures and the potential for Schumpeterian innovation. *Strategic Management Journal*, 19: 1193–1201.

Gardner, H. (1994). The creators' patterns. In D.H. Feldman, M. Csikszentmihalyi and H. Gardner (eds), *Changing the World: A Framework of the Study of Creativity*. Westport: Praeger, pp. 69–70.

Garrick, J. and Clegg, S. (2000). Knowledge work and the new demands of learning. *Journal of Knowledge Management*, 4 (4): 279–86.

Garsten, C. (1999). Betwixt and between: temporary employees as liminal subjects in flexible organizations. *Organization Studies*, 20 (4): 601–17.

Garud, R., Jain, S. and Kumaraswamy, A. (2002). Institutional entrepreneurship in he sponsorship of common technological standards: the case of Sun Microsystems and Java. *Academy of Management Journal*, 45 (1): 196–214.

Garud, R. and Rappa, M.A. (1994). A socio-cognitive model of technology evolution: The case of cochlear implants. *Organization Science*, 5 (3): 344–62.

Gatens, M. (1996). *Imaginary Bodies: Ethics, Power, and Corporeality*. London and New York: Routledge.

Gedo, J.E. and Gedo, M.M. (1992). *Perspectives on Creativity: The Biographical Method*. Norwood, NJ: Ablex.

Gennep, A. van (1960). *The Rites of Passage*. London and New York: Routledge.

Gephart, R.P. Jr. (1996). Postmodernism and the future history of management. *Journal of Management History*, (2) 3: 90–6.

Gherardi, S. (2000). Practice-based theorizing on learning and knowing in organizations. *Organization*, 7 (2), 211–23.

Gherardi, S. and Nicolini, D. (2001). The sociological foundations of organizational learning. In M. Dierkes, A. Berthon, J. Child and I. Nonaka (eds), *Handbook of Organizational Learning & Knowledge*. Oxford: Oxford University Press.

Gibson, C. and Vermeulen, F. (2003). A healthy divide: subgroups as a stimulus for team learning behavior. *Administrative Science Quarterly*, 48: 202–39.

Giddens, A. (1990). *The Consequences of Modernity*. Cambridge: Polity Press.

Gioia, D.A. (1995). Contrasts and convergences in creativity. In C.M. Ford and D.A. Gioia (eds), *Creative Action in Organizations: Ivory Tower Visions and Real World Voices*. Thousand Oaks: Sage, pp. 317–29.

Gioia, D.A. and Chittipeddi, K. (1991). Sensemaking and sensegiving in strategic change initiation. *Strategic Management Journal*, 12: 433–48.

Girard, R. (1977). *Violence and the Sacred*, trans. Patrick Gregory. Baltimore and London: John Hopkins University Press.

Glynn, M.A. (1996). Innovative genius: a framework for relating individual and organizational intelligence to innovation. *Academy of Management Review*, 21: 1081–1111.

Goffman, E. (1959). *The Presentation of Self in Everyday Life*. New York: Doubleday Anchor.

Goodman, R.A. and Sproull, L.S. (1990). *Technology and Organization*. San Francisco: Jossey-Bass.

Goodman, M. (1995). *Creative Management*. Hertfordshire: Prentice Hall.

Gourlay, S. (2004). Knowing as semiosis: steps toward a reconceptualization of 'tacit knowledge'. In H. Tsoukas and N. Mylonopoulos (eds), *Organization Knowledge Systems: Knowledge, Learning and Dynamic Capabilities*. Basingstoke and New York: Palgrave, pp. 86–104.

Grant, R.M. (1996). Prospering in dynamically-competitive environments: organizational capability as knowledge integration. *Organizational Science*, 20 (4): 375–87.

Gregg, S.M. (1997). An overview of the pharmaceutical industry. *Occupational Medicine*, State of the Arts Reviews, January–March, 12 (1). Philadelphia: Hanley & Belfus.

Greve, H.R. (1998). Managerial cognition and the mimetic adoption of market positions: What you see is what you do. *Strategic Management Journal*, 19: 967–88.

Greve, H.R. and Taylor, A. (2000). Innovations as catalysts for organizational change: shifts in organizational cognition and search. *Administrative Science Quarterly*, 45: 54–80.

Griffin, A.E.C., Colella, A. and Goparaju, S. (2000). Newcomer and organizational socialization tactics: an interactionist perspective. *Human Resource Management Review*, 10 (4): 453–74.

Grint, K. (1997). *Fuzzy Management: Contemporary Ideas and Practices at Work*. Oxford: Oxford University Press.

Grint, K. and Woolgar, S. (1997). *The Machine at Work: Technology, Work and Organization*. Cambridge: Polity Press.

Griseri, P. (2002). *Management Knowledge: A Critical View*. Basingstoke: Palgrave.

Grosz, E. (1994). *Volatile Bodies: Toward a Corporeal Feminism*. Bloomington and Indianapolis: Indiana University Press.

Grosz, E. (1995). *Space, Time, and Perversion: Essays in the Politics of Bodies*. New York and London: Routledge.

Grosz, E. (2001). *Architecture From the Outside: Essays on Virtual and Real Spaces*. Cambridge: The MIT Press.

Gruber, H. (1981). *Darwin on Man: A Psychological Study of Scientific Creativity*, 2nd edn. Chicago: University of Chicago Press.

Gruber, H.E. and Davies, S.N. (1988). Inching our way up Mount Olympus: the evolving system approach to creative thinking. In R.J. Sternberg (ed.), *The Nature of Creativity: Contemporary and Psychological Perspectives*, pp. 243–70.

Gubrium, J.F. and Holstein, J.A. (2003). Postmodern sensibilities. In Jaber F. Gubrium and James A. Holstein (eds), *Postmodern Interviewing*. London, Thousand Oaks and New Delhi: Sage, pp. 3–16.

Guilford, J.P. (1950). Creativity. *American Psychologist*, 5: 444–54.

Guilford, J.P. (1967). *The Nature of Human Intelligence*. New York: McGraw-Hill.

Gunther McGrath, R. (2001). Exploratory learning, innovative capacity, and managerial oversight. *Academy of Management Journal*, 44 (1): 118–31.

Habermas, J. (1968). *Knowledge and Human Interest*. London: Heinemann.

Hacking, I. (1983). *Representing and Intervening: Introductory Topics in the Philosophy of Natural Science*. Cambridge: Cambridge University Press.

Hage, J. and Hollingsworth, J.R. (2000). A strategy for the analysis of idea innovation networks and institutions. *Organization Studies*, 21 (5): 971–1004.

Halliday, R.G. (1999). Profile of the pharmaceutical industry from 1998 to 2000, R&D expenditure and staffing. *CMR International*.

Hara, T. (2003). *Innovation in the Pharmaceutical Industry: The Process of Drug Discovery and Development*. Northampton, MA: Edward Elgar Publications.

Hargadon, A. and Sutton, R.I. (2000). Building an innovation factory. *Harvard Business Review*, 78 (3): 157–66.

Harrington, D.M. (1999). Conditions and settings/environment. In M.A. Runco and S.R. Pritzker (eds), *Encyclopaedia of Creativity*. London: Academic Press, p. 324.

Harvey, A. (2001). A dramaturgical analysis of charismatic leader discourse. *Journal of Organizational Change Management*, 14 (3): 253–65.

Hayden, P. (1998). *Multiplicity and Becoming: The Pluralist Empiricism of Gilles Deleuze*. New York: Peter Lang.

Hayes, N. and Walsham, G. (2003). Knowledge sharing and ICTs: a relational perspective. In M. Easterby-Smith and M.A. Lyles (eds), *Handbook of Organization Learning and Knowledge Management*. Oxford and Malden: Blackwell, pp. 54–77.

Heckscher, C. and Donnellon, A. (1994). *The Postbureaucratic Organization: New Perspectives on Organizational Change*. Thousand Oaks, London and New Delhi: Sage.

Hedberg, B. and Jönsson, S. (1978). Designing semi-confusing information systems for organizations in changing environments. *Accounting, Organizations and Society*, 3 (1): 47–64.

Heidegger, M. (1977). *The Question Concerning Technology and Other Essays*. New York: Harper and Row.

Henderson, R. and Cockburn, I. (1994). Measuring competence? Exploring firm effects in pharmaceutical research. *Strategic Management Journal*, 15: 63–84.

Henry, J. (2001). *Creativity and Management Perceptions*. Thousand Oaks, CA: Sage.

Horrobin, D.F. (2000). Innovation in the pharmaceutical industry. *Journal of the Royal Society of Medicine*, 93: 341–5.

Horrobin, D.F. (2002). Effective clinical innovation: an ethical imperative. *Lancet*, 359: 1857–1858.

Hitt, M.A., Ireland, D., Camp, S.M. and Sexton, D.L. (2001). Guest editor's introduction to the special issue Strategic entrepreneurship. Entrepreneurship strategies for wealth creation. *Strategic Management Journal*, 22: 479–91.

Hjort, D. (1999). HLPE: On HRM and entrepreneurship. *Organization*, 6 (2): 307–24.

Hjort, D. (2003). *Rewriting Entrepreneurship: For a New Perspective on Organizational Creativity*, Malmö: Liber; Oslo: Abstrakt; Copenhagen: Copenhagen Business School Press.

Hopkin, K. (2001). The post-genome project. *Scientific American*, August issue.

Howard, K. (2000). The bioinformatics gold rush. *Scientific American*, July issue.

Huber, G.P. (1991). Organizational learning: the contributing processes and the literatures. *Organization Science*, 2 (1): 88–115.

Hughes, D. (1998). Predicting the future for R&D – science or art? *Drug Discovery Today*, (3): 487–89.

Huff, A.S. (ed.) (1992). *Mapping Strategic Thought*. New Jersey: Wiley.

Huizinga, J. (1949). *Homo Ludens: A Study of the Play-Element in Culture*. London: Routledge & Kegan Paul.

Hullman, A. (2000). Generation, transfer and exploitation of new knowledge. In A. Jungmittag, A. Reger and G. Reiss (eds), *Changing Innovation in the Pharmaceutical Industry – Globalization and New Ways of Drug Development*. Berlin: Springer.

Humphreys, M. and Brown, A.D. (2002). Narratives of organizational identity and identification: a case study of hegemony and resistance. *Organization Studies*, 23 (3): 421–47.

Hurley, J. (2003). *Scientific Research Effectiveness*: The Organisational Dimension. Boston: Kluwer Academic Publishers.

Huy, Q.N. (1999). Emotional capability, emotional intelligence, and radical change. *Academy of Management Review*, 24 (2): 325–45.

Huzzard, T. and Östergren, K. (2002). When norms collide: learning under organizational hypocrisy. *British Journal of Management*, 13: 47–59.

Isaksen, S.G. (1987). *Frontiers of Creativity Research: Beyond the Basics*. Buffalo, NY: Bearly.

Isaksen, S.G., Lauer, K.J., Ekvall, G. and Britz, A. (2001). Perceptions of the best and worst climate for creativity: preliminary validation evidence for the situational outlook questionnaire. *Creativity Research Journal*, 13 (2): 171–84.

Jackson, P. (1999). *Virtual Working: Social and Organizational Dynamics*. London: Routledge.

Jacob, N. (1998). *Creativity in Organisations*. New Delhi: Wheeler Publishing.

Jacques, R. (1996). *Manufacturing the Employee*. London: Sage.

Jain, K.K. (2000). Transforming innovation and commercialization in drug discovery. *Drug Discovery Today*, 5 (8): 318–20.

Janesick, V. (2000). The choreography of qualitative design. In N.K. Denzin and Y.S. Lincoln (eds), *Handbook of Qualitative Research*, 2nd edn. London, Thousand Oaks: Sage, pp. 379–99.

Jasanoff, Sheila, Markle, Gerald E., Peterman, James C. and Pinch, Trevor (eds) (1995), *Handbook of Science and Technology Studies*. Thousand Oaks, London and New Delhi: Sage.

Jassawalla, A.R. and Sashittal, Hemant C. (2002). Cultures that support product innovation processes. *Academy of Management Executive*, 16 (3): 42–54.

Jeffcut, P. (2000). Management and the creative industries. *Studies in Cultures, Organizations and Societies*, 6: 123–7.

Johnson, G. (1996). What place for R&D in tomorrow's drug industry? *Drug Discovery Today*, 1 (3): 117–121.

Jones, E. (2001). *The Business of Medicine*. London: Profile Books.

Jung, D.I. (2001). Transformational and transactional leadership and their effects on creativity in groups. *Creativity Research Journal*, 13: 185–97.

Kakabadse, A. (1984). Politics of a process consultant. In A. Kakabadse and C. Parker (eds), *Power, Politics and Organizations*. Chichester: Wiley, pp. 169–83.

Kalling, T. and Styhre, A. (2003). *Knowledge Sharing in Organizations*. Malmö: Liber; Oslo: Abstrakt; Copenhagen: Copenhagen Business School Press.

Kallinikos, J. (1996). *Organizations in an Age of Information*. Lund: Academia Adacta.

Kallinikos, J. (1998). Organized complexity: posthumanist remarks on the technologizing of intelligence. *Organization*, 5 (3): 371–96.

Kamoche, K. and Pina C.M. (2001). Minimal structures: from jazz improvization to product innovation. *Organization Studies*, 22 (5): 733–64.

Kao, J. (1996). *Jamming: The Art and Discipline of Business Creativity*. London: Harper Collins.

Katz, D. and Kahn, R.L. (1966). *The Social Psychology of Organizations*. New York: Wiley.

Kazanjian, R.K., Drazin, R. and Glynn, M. (2000). Creativity and technological learning: the roles of organization architecture and crisis in large-scale projects. *Journal of Engineering Technology Management*, 17: 273–98.

Kegan, P. and Pickering, A. (1995). *The Mangle of Practice: Time, Agency, and Science*. Chicago and London: The University of Chicago Press.

Kelemen, M. (2001). Discipline at work: distal and proximal views. *Studies in Cultures, Organizations and Societies*, 7: 1–23.

Kingdon, J.W. (1984). *Agendas, Alternatives and Public Policies*. Boston and Toronto: Little, Brown and Company.

Kidwell, R.E. and Bennett, N. (1994). Employee reactions to electronic control systems. *Group & Organization Management*, 19 (2): 203–19.

Kilbourne, L.M. and Woodman, R.W. (1999). Barriers to organizational creativity. In R.E. Purser and A. Montuori (eds), *Social Creativity*, vol. II. Cresskill, NJ: Hampton Press, pp. 125–50.

Kirzner, I.M. (1973). *Competition and Entrepreneurship*. Chicago: University of Chicago Press.

Kitchener, M. (2000). The 'bureaucratization' of professional roles: the case of clinical directors in UK hospitals. *Organization*, 7 (1): 129–54.

Kitchener, M. (2002). Mobilizing the logic of managerialism in professional fields: the case of academic health centre mergers. *Organization Studies*, 23 (3): 391–420.

Knight, F.H. (1971). *Risk, Uncertainty and Profit*. Chicago: University of Chicago Press.

Knorr Cetina, K. (1995). How superorganisms change: consensus formation and the social ontology of high-energy physics experiments, *Social Studies of Science*, 25: 119–147.

Knorr Cetina, K. (1999). *Epistemic Cultures: How Sciences Make Knowledge*. Cambridge: Harvard University Press.

Koh, A. (2000). Linking learning, knowledge creation, and business creativity: a preliminary assessment of the East Asian quest for Creativity. *Technological Forecasting and Social Change*, 64: 85–100.

Kolakowski, L. (1985). *Bergson*. Oxford and New York: Oxford University Press.

Korczynski, M. (2000). The political economy of trust. *Journal of Management Studies*, 37 (1): 1–21.

Koretz, S. and Lee, G. (1998). Knowledge management and drug development. *Journal of Knowledge Management*, 2 (2): 53–8.

Kostera, M. (1997). The kitsch-organization. *Studies in Culture, Organization & Societies*, 3: 163–77.

Kotter, J.P. (1987). *The Leadership Factor*. New York: The Free Press.

Kreiner, K. (1995). In search of relevance: project management in drifting organizations. *Scandinavian Journal of Management*, 11 (4): 335–46.

Kreiner, K. and Mouritsen, J. (2003). Knowledge management as technology: making knowledge manageable. In B. Czarniawska and G. Sevón (eds), *The Northern Lights: Organization Theory in Scandinavia*. Malmö: Liber; Oslo: Abstrakt; Copenhagen: Copenhagen Business School Press.

Kuhn, T.S. (1962/1996). *The Structure of Scientific Revolutions*, 3rd edn. Chicago: University of Chicago Press.

Kundera, M. (1988). *The Art of the Novel*, trans. L. Asher. New York: Grove Press.

Kurland, Nancy B. and Pelled, Lisa Hope, (2000). Passing the word: toward a model of gossip and power in the workplace. *Academy of Management Review*, 25 (2): 428–38.

Kvale, S. (1996). *InterViewing*. London: Sage.

Lacetera, N. and Orsenigo, L. (2001). Political regimes, technological regimes and innovation in the evolution of the pharmaceutical industry in the USA and in Europe. Conference Paper, Conference on Evolutionary Economics, Johns Hopkins University, Baltimore, 30–1 March.

Lakatos, I. (1970). Falsification and the methodology of scientific research programmes. In I. Lakatos and A. Musgrave (eds), *Criticism and the Growth of Knowledge*. Cambridge: Cambridge University Press.

Lanzara, G.F. and Patriotta, G. (2001). Technology and the courtroom: an inquiry into knowledge making in organizations. *Journal of Management Studies*, 38 (7): 943–71.

Lasswell, H.D. (1950 [1936]). *Politics: Who Gets What, When, How*. New York: P. Smith.

Latour, B. (1987). *Science in Action*. Cambridge: Harvard University Press.

Latour, B. (1991). Technology is society made durable. In J. Law (ed.), *A Sociology of Monsters: Essays on Power, Technology and Domination*. London and New York: Routledge.

Latour, B. and Woolgar, S. (1979). *Laboratory Life: The Construction of Scientific Facts*. Princeton, NJ: Princeton University Press.

Law, J. (2002). *Aircraft Stories: Decentering the Object in Technoscience*. Durham: Duke University Press.

Leonard, D. and Sensiper, S. (1998). The role of tacit knowledge in group innovation. *California Management Review*, 40 (3): 112–30.

Leonard-Barton, D. (1988). Implementation as mutual adoption of technology and organization. *Research Policy*, 17: 251–67.

Leonard-Barton, D. (1995). *Wellspring of Knowledge: Building and Sustaining the Sources of Innovation*. Boston: Harvard Business School Press.

Lesko, J.L., Rowland, M., Peck, C.C. and Blaschke T.F. (2000). Optimizing the science of drug development: opportunities for better candidate selection and accelerated evaluation in humans. *Pharmaceutical Research*, 17 (11): 1335–1341.

Levinthal, D.A. and March, J.G. (1993). The myopia of learning. *Strategic Management Journal*, 14: 95–112.

Lincoln, Y.S. and Guba, E.G. (2000). Paradigmatic controversies, contradictions, and emerging confluences. In N.K. Denzin and Y.S. Lincoln (eds), *Handbook of Qualitative Research*. Thousand Oaks, CA: Sage, pp. 377–92.

Linstead, S. (2000). Organizational kitsch. *Organization*, 9 (4): 657–82.

Linstead, S. (2002). Organization as reply: Henri Bergson and causal organization theory. *Organization*, 9 (1): 95–111.

Little, S., Quintas, P. and Ray, T. (2002). *Managing Knowledge: An Essential Reader*. London, Thousand Oaks and New Delhi: Sage.

Llewellyn, S. (2001). 'Two-way windows'. Clinicians as medical managers. *Organization Studies*, 22 (4): 593–623.

Locke, E.A. and Kirkpatrick, S.A. (1995). Promoting creativity in organizations. In C. Ford and D.A. Gioia (eds), *Creative Action in Organizations: Ivory Tower Visions and Real World Voices*. Thousand Oaks: Sage Publications, pp. 115–20.

Lucas, G.R. (1989). *The Rehabilitation of Whitehead: An Analytic and Historical Assessment of Process Philosophy*. Albany: State of New York University Press.

Luhmann, N. (1982). *The Differentiation of Society*. New York: Columbia University Press.

Luhmann, N. (1995). *Social Systems*. Stanford: Stanford University Press.

Lundin, R.A. and Söderholm, A. (1995). A theory of the temporary organization, *Scandinavian Journal of Management*, 11 (4): 437–55.

Lynch, M. (1985). *Art and Artifact in Laboratory Science: A Study of Shop Work and Shop Talk in a Research Laboratory*. London: Routledge & Kegan Paul.

Lynch, M. (2002). Protocols, practices and the reproduction of technique in molecular biology. *British Journal of Sociology*, 53 (2): 203–20.

Lyotard, J.-F. (1979). *The Postmodern Condition: A Report on Knowledge.* Manchester: Manchester University Press.

MacKenzie, Donald (1999). Slaying the kraken: the sociohistory of a mathematical proof. *Social Studies of Science,* 29 (1): 7–60.

Madjar, N., Oldham, G.R. and Pratt, M.G. (2002). There's no place like home? The contributions of work and nonwork creativity support to employees' creative performance. *Academy of Management Journal,* 45 (4): 757–67.

Magyari-Beck, I. (1999). Creatology. In M.A. Runco and S.R. Pritzker (eds), *Encyclopaedia of Creativity.* London: Academic Press, pp. 433–41.

March, J.G. (1991). Exploration and exploitation in organizational learning. *Organization Science,* 2 (1): 71–87.

March, J.G. and Olsen, J. (1976). *Ambiguity and Choice in Organizations.* Oslo: Universitetsforlaget.

March, J.G. and Simon, H.A. (1958). *Organizations,* 2nd edn. Oxford: Blackwell.

Marks, J. (1998). *Gilles Deleuze: Vitalism and Multiplicity.* London: Pluto Press.

Massumi, B. (2002). *Parables of the Virtual: Movement, Affect, Sensation.* Durham and London: Duke University Press.

Maturana, H.R. and Varela, F.J. (1992). *The Tree of Knowledge: The Biological Roots of Human Understanding.* Boston and London: Shambala.

May, C. (2002). *The Information Society: A Sceptical View.* Cambridge: Polity Press.

Mayer, R.E. (1999). Fifty years of creativity research. In R.J. Sternberg (ed.), *Handbook of Creativity.* Cambridge, UK: Cambridge University Press, pp. 449–60.

McFadzean, E. (2000). What can we learn from creative people? The story of Brian Eno. *Management Decision,* 38 (1): 51–6.

McGail, B.A. (2002). Confronting electronic surveillance: desiring and resisting new technologies. In S. Woolgar, (ed.), *Virtual Society? Technology, Cyberbole, Reality.* Oxford and New York: Oxford University Press.

McGregor, D. (1960). *The Human Side of Enterprise.* New York: McGraw-Hill.

McLuhan, M. (1962). *The Gutenberg Galaxy: The Making of Typographic Man.* London: Routledge & Kegan Paul.

Michael, W.P. (1999). Guilford's view. In M.A. Runco and S.R. Pritzker (eds), *Encyclopaedia of Creativity.* London: Academic Press, pp. 785–97.

Michaels, M. (2000). *Reconnecting Culture, Technology and Nature: From Society to Heterogeneity.* London and New York: Routledge.

Miller, W.C. (1999). *Flash of Brilliance: Inspiring Creativity Where You Work.* New York: Perseus Books.

Mintzberg, H. (1983). *Structure In Fives.* Englewood Cliffs: Prentice Hall.

Mitchell, R.K. (1996). Oral history and expert scripts: demystifying the entrepreneurial experience. *Journal of Management History,* 2 (3): 50–67.

Moore, F.C.T. (1996). *Bergson: Thinking Backwards.* Cambridge: Cambridge University Press.

Montouri, A. (1992). Two books on creativity. *Creativity Research Journal,* 5: 199–203.

Morris, T. (2001). Asserting property rights: knowledge codifications in the professional service firm. *Human Relations,* 54 (7): 819–38.

Mouritsen, J., Larsen, H.T. and Bukh, P.N.D. (2001). Intellectual capital and the 'capable firm': Narrating, visualising and numbering for management knowledge, *Accounting, Organization and Society,* 26: 735–62.

Mueller, F. and Dyerson, R. (1999). Expert humans or expert organizations. *Organization Studies,* 20 (2): 225–56.

Mullarkey, J. (1999). *Bergson and Philosophy*. Edinburgh: Edinburgh University Press.

Mumford, L. (1934). *Technics and Civilization*, San Diego, New York and London: Harcourt Brace Jovanovich.

Mumford, M.D. (2000). Managing creative people: strategies and tactics for innovation. *Human Resource Management Review*, 10 (3): 313–51.

Mumford, M.D. and Connely, M.S. (1999). Leadership. In M.A. Runco and S.R. Pritzker (eds), *Encyclopaedia of Creativity*. London: Academic Press, pp. 139–45.

Mumford, M.D., Scott, G.M., Gaddis, B. and Strange, J.M. (2002). Leading creative people: orchestrating expertise and relationships. *The Leadership Quarterly*, 13: 705–50.

Naurin, D. (2004). *Dressed for Politics: Why Increasing Transparency in the European Union Will Not Make Lobbyists Behave Any Better Than They Already Do*, PhD Diss., Dept. of Political Science, Gothenburg University.

Nelson, R.R. (2003). On the uneven evolution of human know-how. *Research Policy*, 32: 909–22.

Nevil, E.C., DiBella, A.J. and Gould, J.M. (1995). Understanding organizations as learning systems. *Sloan Management Review*, Winter: 73–85.

Newell, S., Scarborough, H., Swan, J. and Hislop, D. (2000). Intranets and knowledge management: de-centred technologies and the limits of technological discourse. In Craig Prichard, Richard Hull, Mike Chumer and Hugh Willmott (eds), *Managing Knowledge: Critical Investigations of Work and Learning*. New York: St. Martin's Press.

Nickerson, R.S. (1999). Enhancing creativity. In R.J. Sternberg (ed.), *Handbook of Creativity*. Edinburgh: Cambridge University Press, pp. 392–430.

Nightingale, P. (1998). A cognitive model of innovation. *Research Policy*, 27: 689–709.

Nightingale, P. (2003). If Nelson and Winter are only half right about tacit knowledge, which half? A Searlean critique of codification. *Industrial and Corporate Change*, 12 (2): 149–83.

Nobel, R. and Birkinshaw, J. (1998). Innovation in multinational corporations: control and communication patterns in international R&D operations. *Strategic Management Journal*, 19: 479–96.

Nonaka, I. and Takeushi, H. (1995). *The Knowledge-Creating Company*. Oxford: Oxford University Press.

Noon, M., Jenkins, S., and Martinez L.M. (2000). Fads, techniques and control: the competing agendas of TPM and TECEX at the Royal Mail (UK). *Journal of Management Studies*, 37 (4): 499–520.

Oberholzer-Gee, F. and Inamdar, S.N. (2004). Merck's recall of Rofecoxib – a strategic perspective. *The New England Journal of Medicine*, 351 (21): 2147–2150.

Ogbor, J.O. (2000). Mythicizing and reification in entrepreneurial discourse: Ideology-critique of entrepreneurial studies. *Journal of Management Studies*, 37 (5): 605–35.

Oldham, G.R. and Cummings, A. (1996). Employee creativity: personal and contextual factors at work. *Academy of Management Review*, 39 (3): 607–34.

Olkowski, D. (1999). *Gilles Deleuze and the Ruin of Representation*. Berkeley: University of California Press.

Orlikowski, W.J. (1992). The duality of technology: rethinking the concept of technology in organizations. *Organization Science*, 3 (3): 398–427.

Orlikowski, W.J. (2000). Using technology and constituting structures: a practice lens for studying technology in organizations. *Organization Science*, 11 (4): 404–28.

Orlikowski, W.J. (2002). Knowing in practice: enacting a collective capability in distributed organizing. *Organization Science*, 13 (3): 249–73.

Orr, J.E. (1996). *Talking About Machines: An Ethnography of a Modern Job*. Ithaca and London: Cornell University Press.

Osbourne, T. (2003). Against 'creativity': a philistine rant. *Economy and Society*, 32 (4): 507–25.

Oseen, C. (1997). Luce Irigaray, Sexual difference and theorizing leaders and leadership. *Gender, Work and Organization*, 4 (3): 170–84.

O'Shea, A. (2002). The (r)evolution of new product innovation. *Organization*, 9 (1): 113–25.

Parker, M. (2000). *Organization, Culture and Identity: Unity and Division at Work*. London, Thousand Oaks and New Delhi: Sage.

Parnell, J.A., Lester, D.L., and Menefee, M.L. (2000). Strategy as a response to organizational uncertainty: an alternative perspective on the strength–performance relationship. *Management Decision*, 38: 301–92.

Patriotta, G. (2003). Sensemaking on the shop floor: narratives of knowledge in organizations. *Journal of Management Studies*, 40(2): 349–75.

Patton, P. (2001). Notes for a glossary. In Gary Genosko (ed.), *Deleuze and Guattari: Critical Assessment of Leading Philosophers*. London and New York: Routledge.

Pawlowsky, P., Forslin, J. and Reinhardt, R. (2001). Practices and tools of organizational learning, in M. Dierkes, A. Berthon, J. Child and I. Nonaka (eds), *Handbook of Organizational Learning & Knowledge*. Oxford: Oxford University Press.

Payne, S.L. (1996). Qualitative research and reflexive faculty change potentials. *Journal of Organizational Change Management*, 9 (2): 20–31.

Peirce, C.S. (1991). *Peirce on Signs: Writings on Semiotics by Charles Sanders Peirce*, J. Hopes, E., Chapel Hill & London: The University of North Carolina Press.

Penrose, E.T. (1959). *The Theory of the Growth of the Firm*. Oxford: Blackwell.

Pentland, B. (1999). Building process theory with narrative from description and explanation. *Academy of Management Review*, 24 (4): 711–24.

Peteraf, M. (1993). The cornerstone of competitive advantage. *Strategic Management Journal*, 14: 179–91.

Pettigrew, A.M. (1973). *The Politics of Organizational Decision-Making*. London: Tavistock.

Pfeffer, J. and Sutton, R.I. (1999). *The Knowing–Doing Gap: How Smart Companies Turn Knowledge into Action*. Cambridge: Harvard University Press.

Pickering, A. (1995). *The Mangle of Practice: Time, Agency, and Science*. Chicago and London: The University of Chicago Press.

Pisano, G.P. (1997). *The Development Factory: Unlocking the Potential of Process Innovation*. Boston: Harvard Business School Press.

Plucker, J.A. and Renzulli, J.S. (1999). Psychometric approaches to the study of human creativity. In R.J. Sternberg (ed.), *Handbook of Creativity*. Edinburgh: Cambridge University Press, pp. 35–61.

Plucker, J.A. and Runco, M.A. (1998). The death of creativity measurement has been greatly exaggerated: current issues, recent advances, and future directions in creativity assessment. *Roeper Review*, 21 (1): 36–9.

Polanyi, M. (1958). *Personal Knowledge: Toward a Post-Critical Philosophy.* Chicago: Chicago University Press.

Policastro, E. (1995). Creative intuition: an integrative review. *Creative Research Journal,* 8: 99–113.

Policastro, E. (1999). Intuition. In M.A. Runco and S.R. Pritzker (eds), *Encyclopaedia of Creativity.* London: Academic Press, pp. 89–91.

Polkinghorne, D.E. (1988). *Narrative Knowing and the Human Sciences.* Albany: State University of New York Press.

Popper, K.R. (1963). *Conjectures and Refutations: The Growth of Scientific Knowledge.* London and Henley: Routledge and Kegan Paul.

Poster, M. (2001). *What's the Matter with the Internet?* Minneapolis: University of Minnesota Press.

Power, M. (2004). Counting, control and calculation: reflection on measuring and management. *Human Relations,* 57 (6): 765–83.

Powell, W.W. and Smith-Doerr, L. (1996). Interorganizational collaboration and the locus of innovation: networks of learning in biotechnology. *Administrative Science Quarterly,* 41: 116–45.

Pritchard, C. (2000). Know, learn, share! The knowledge phenomena and the construction of a consumptive-communicative body. In C.P.R. Hull, M. Chumer and H. Willmott (eds), *Managing Knowledge: Critical Investigations of Work and Learning.* New York: St Martin's Press.

Rabinow, P. (1996). *Making PCR: A Story of Biotechnology.* Chicago and London: The University of Chicago Press.

Rabinow, P. (1999). *French DNA: Trouble in Purgatory.* Chicago and London: The University of Chicago Press.

Ready, D.A. (2002). How storytelling builds the next generation leaders. *Sloan Management Review,* Summer: 63–9.

Reed, M.I. (1996). Expert power in late modernity: an empirical review and theoretical synthesis. *Organization Studies,* 17 (4): 573–97.

Reiss, T. and Hinze, S. (2000). Innovation process and techno-scientific dynamics. In A. Jungmittag, A. Reger and G. Reiss (eds), *Changing Innovation in the Pharmaceutical Industry – Globalization and New Ways of Drug Development.* Berlin: Springer, pp. 53–69.

Richards, T. and Moger, S. (2000). Creative leadership processes in project team development: an alternative to Tuckman's stage model. *British Journal of Management,* 11: 273–83.

Rickards, T. (1991). Innovation and creativity: woods, trees, and pathways. *R&D Management,* 21 (2): 97–108.

Rickards, T. (1999). *Creativity and the Management of Change.* Oxford: Blackwell Business.

Rickards, T. and De Cock, C. (1999). Sociological paradigms and organizational creativity. In R.E. Purser and A. Montuori (eds), *Social Creativity,* vol. II. Cresskill, NJ: Hampton Press, pp. 235–56.

Roberts, R.M. (1989). *Serendipity: Accidental Discoveries in Science.* New York: John Wiley, pp. 75–82.

Roberts, R. (1998). Managing innovation: the pursuit of competitive advantage and the design of innovation intense environments. *Research Policy,* 27: 159–75.

Roberts, P.W. (1999). Product innovation, product–market competititon and persistent profitability in the US pharmaceutical industry. *Strategic Management Journal,* 20: 655–70.

Robertson, M. and Swan, J. (2003). 'Control – what control?' Culture and ambiguity within a knowledge intensive firm. *Journal of Management Studies*, 40 (4): 831–58.

Robinson, A.G. and Stern, S. (1997). *Corporate Creativity*. San Francisco: Berrett-Koehler Publishers.

Ropo, A. and Parviainen, J. (2001). Leadership and bodily knowledge in expert organizations: epistemological rethinking. *Scandinavian Journal of Management*, 17: 1–18.

Rose, P. (2002). *On Whitehead*. Belmont, CA: Wadsworth.

Rouxel, G. (2001). Cognitive–affective determinants of performance in mathematics and verbal domains. *Learning and Individual Differences*, 12: 287–310.

Runco, M.A. (1999). Time. In M.A. Runco and S.R. Pritzker (eds), *Encyclopaedia of Creativity*. London: Academic Press, pp. 659–63.

Runco, M.A., Nemiro, J. and Walberg, H.J. (1998). Personal explicit theories of creativity. *Journal of Creative Behaviour*, 32: 1–17.

Runco, M.A. and Sakamoto, S. O. (1999). Experimental studies of creativity. In R.J. Sternberg (ed.), *Handbook of Creativity*. Cambridge, England: Cambridge University Press, pp. 62–92.

Ryan, R.M. (1982). Control and information in the intrapersonal sphere: an extension of cognitive evaluation theory. *Journal of Personality and Social Psychology*, 43: 450–61.

Salaman, G. and Storey, J. (2002). Managers' theories about the process of innovation. *Journal of Management Studies*, 39 (2): 147–65.

Santos, F. (2003). The co-evolution of firms and their knowledge environment: insights from the pharmaceutical industry. *Technological Forecasting & Social Change*, 70: 687–715.

Saviotti, P.P. (1998). On the dynamics of appropriability, of tacit and of codified knowledge. *Research Policy*, 26: 843–56.

Schmid, E.F. and Smith, D.A. (2002a). Discovery, innovation and the cyclical nature of the pharmaceutical business. *Drug Discovery Today*, 7 (10): 563–8.

Schmid, E.F. and Smith, D.A. (2002b). Should scientific innovation be managed? *Drug Discovery Today*, 7 (18): 941–5.

Schneider, M. and Teske, P. (1992). Toward a theory of the political entrepreneur: evidence from local government. *American Political Science Review*, 86 (3): 737–47.

Schoenfeldt, L.F. and Jansen, K.J. (1997). Methodological requirements for studying creativity in organizations. *Journal of Creative Behavior*, 31 (1): 73–91.

Schultze, U. and Orlikowski, W.J. (2001). Metaphors of virtuality: shaping an emergent reality. *Information and Organization*, 11: 45–77.

Schumpeter, J.A. (1942). *Capitalism, Socialism, and Democracy*. New York: Harper & Row.

Scott, S.G. and Bruce, R.A. (1994). Determinants of innovative behavior – A path model of individual innovation in the workplace. *Academy of Management Journal*, 37 (3): 580–607.

Selen, W. (2000). Knowledge management in resource-based competitive environments: a roadmap for building learning organizations. *Journal of Knowledge Management*, 4 (4): 346–53.

Selznick, P. (1957). *Leadership in Administration*. Berkeley: University of California Press.

Serres, M. (1995). *Genesis*. Ann Arbor: University of Michigan Press.

Sewell, G. (1998). The discipline of teams: the control of team-based industrial work through electronic and peer surveillance. *Administrative Science Quarterly*, 43: 397–428.

Shalley, C.E. and Perry-Smith, J.E. (2001). Effects of social-psychological factors on creative performance: the role of informational and controlling expected evaluation and modeling experience. *Organisational Behaviour and Human Decision Processes*, 84 (1): 1–22.

Shani, A.B. and Lau, J.B. (2000). *Behaviour in Organizations: An Experimental Approach*. Boston: McGraw-Hill.

Shani, A.B., David, A. and Wilson, C. (2003). Collaborative research. In N. Adler, A.B. Shani and A. Styhre (eds), *Collaborative Research in Organizations*. Thousand Oaks, CA: Sage, pp. 83–100.

Sherburne, D.W. (1966). *A Key to Whitehead's Process and Reality*. Chicago: Chicago University Press.

Shin, S.J. and Zhou, Jing (2003). Transformational leadership, conservation, and creativity: evidence from Korea. *Academy of Management Journal*, 46 (6): 703–14.

Shotter, J. and Billig, M. (1998). A Bakhtinian psychology: from out of the heads of the individuals and into the dialogues between them. In M.M. Bell and M. Gardiner (eds), *Bakhtin and the Human Sciences*. London, Thousand Oaks and New Delhi: Sage.

Siegel, S. and Kaemmerer, W. (1978). Measuring the perceived support for innovation in organizations. *Journal of Applied Psychology*, 63: 553–62.

Silverman, D. (1993). *Interpreting Qualitative Data*. London, Thousand Oaks and New Delhi: Sage.

Simon, H.A. (1957). *Models of Man*. New York: Wiley.

Simonton, D.K. (1984). *Genius Creativity & Leadership: Historiometric Inquiries*. Cambridge, MA: Harvard University Press.

Slywotzky, A.J. (1996). *Value Migration: How to Think Several Moves Ahead of the Competition*. Boston: Harvard Business School Press.

Spender, J.C. (1996). Making knowledge the basis of a dynamic theory of the firm, *Strategic Management Journal*, 17, Winter Special Issue: 45–62.

Spinoza, B. (1994). *Ethics*. London: Penguin.

Stacey, R. (1996). *Complexity and Creativity in Organizations*. San Fransisco: Berrett Koehler Publishers.

Starbuck, W.H. (1992). Learning by knowledge-intensive firms. *Journal of Management Studies*, 29 (6): 713–40.

Starbuck, W.H. (2003). The origins of organization theory. In H. Tsoukas and C. Knudsen (eds) (2003), *The Oxford Handbook of Organization Theory: Meta-Theoretical Perspectives*. Oxford and New York: Oxford University Press.

Staw, B. (1995). Why no one really wants creativity. In C.M. Ford and D.A. Gioia (eds), *Creative Action in Organizations: Ivory Tower Visions and Real World Voices*. Thousand Oaks: Sage Publications, pp. 161–6.

Sternberg, R.J. (2003). WICS: a model of leadership in organizations. *Academy of Management Learning & Education*, 2 (4): 386–401.

Sternberg R.J. and Lubart, T.I. (1999). The concept of creativity: prospects and Paradigms. In *Handbook of Creativity*. Cambridge, UK: Cambridge University Press, pp. 3–15.

Stewart, T.A. (1997). *Intellectual Capital: The New Wealth of Organizations*. New York: Doubleday.

Steyrer, J. (1998). Charisma and the archetypes of leadership. *Organization Studies*, 19 (5): 807–28.

Stivale, C.J. (1998). *The Two-Fold Thought of Deleuze and Guattari: Intersections and Animations*. New York: The Guilford Press.

Strategic Management Journal (2001). 22, Special issue on Strategic entrepreneurship. Entrepreneurship strategies for wealth creation.

Strati, A. (1999). *Organization and Aethstics*. London, Thousand Oaks and New Delhi: Sage.

Strauss, A.L. and Corbin, J. (1990). *Basics of Qualitative Research*. London, Thousand Oaks and New Delhi: Sage.

Strauss, A.L. and Corbin, J. (1994). Grounded theory methodology: an overview. In N.K. Denzin and Y.S. Lincoln (eds), *Handbook of Qualitative Research*. London: Sage.

Styhre, A. (2003). *Understanding Knowledge Management: Critical and Postmodern Perspectives*. Copenhagen: Copenhagen Business School Press.

Styhre, A., Kohn, K. and Sundgren, M. (2002). Action research as theoretical practice. *Concepts and Transformation*, 7 (1): 93–105.

Styhre, A. and Sundgren, M. (2003a). Creativity as connectivity: a rhizome model of creativity. *International Journal of Internet and Enterprise Management*, 1 (4): 421–36.

Styhre, A. and Sundgren, M. (2003b). 'Management is evil': management control, technoscience and *saudade* in pharmaceutical research. *The Leadership & Organization Development Journal*, 24 (8): 436–46.

Styhre, A. and Sundgren, M. (2005). Action research as experimentation. *Systemic Practice and Action Research*, 18 (1).

Subramaniam, M. and Venkatraman, N. (2001). Determinants of transnational new product development capability: testing the influence of transferring and deploying new knowledge. *Strategic Management Journal*, 22: 359–78.

Sundgren, M. (2003). Whitehead's 'Science and the Modern World' and the concept of organizational creativity – is there a relevance today? Conference paper presented at the EGOS Colloquium (European Group for Organization Studies), Copenhagen, July.

Sundgren, M. (2004). *New Thinking, Management Control, & Instrumental Rationality: Managing Organizational Creativity in Pharmaceutical R&D*, PhD Diss., Chalmers University of Technology, Göteborg.

Sundgren, M. and Styhre, A. (2003a). Creativity a volatile key of success: creativity in new drug development. *Creativity and Innovation Management*, 12 (3): 145–61.

Sundgren, M. and Styhre A. (2003b) Managing organizational creativity. In N. Adler, A.B. Shani and A. Styhre (eds), *Collaborative Research in Organizations*. Thousand Oaks: Sage, pp. 237–53.

Sundgren, M. and Styhre, A. (2004). Intuition and pharmaceutical research: the case of AstraZeneca. *European Journal of Innovation Management*, 7 (4): 267–79.

Sundgren, M., Dimenäs, E., Gustafsson, J.-E. and Selart, M. (2005a). Drivers of organizational creativity: a path model of creative climate in pharmaceutical R&D. *R&D Management*.

Sundgren, M., Selart, M., Ingelgård, A. and Bengtson, C. (2005b). Dialogue-based evaluation as a creative climate indicator: evidence from the pharma industry. *Creativity and Innovation Management*, 14 (1): 84–98.

Surin, K. (1997). The 'epochality' of Deleuzian thought. *Theory, Culture & Society*, 14 (2): 9–21.

Taggar, S. (2002). Individual creativity and group ability to utilize individual creative resources: a multilevel model. *Academy of Management Journal*, 45 (2): 315–32.

Taylor, D. (2003). Fewer new drugs from the pharmaceutical industry. *British Medical Journal*, 326: 408–9.

Teece, D.J. (2000). *Managing Intellectual Capital: Organizational, Strategic and Policy Dimensions*. Oxford and New York: Oxford University Press.

Teece, D.J., Pisano, G. and Shuen, A. (1997). Dynamic capabilities and strategic management. *Strategic Management Journal*, 18 (7): 509–33.

Teubner, G. (2001). Economics of gift – positivity of justice: the mutual paranoia of J. Derrida and N. Luhmann. *Theory, Culture & Society*, 18(1): 29–47.

Thomke, S. (2001). Enlightened experimentation: the new imperative for innovation. *Harvard Business Review*, 79 (2): 66–75.

Thomke, S. and Kuemmerle, W. (2002). Asset accumulation, interdependence and technological change: evidence for the pharmaceutical industry. *Strategic Management Journal*, 23: 619–35.

Thompson, V.A. (1969). *Bureaucracy and Innovation*. University: The University Press of Alabama.

Torrance, E.P. (1974). *Torrance Test of Creative Thinking: Norms-technical Manual*. Lexington, MA: Ginn.

Touraine, A. (1971). *The Post-Industrial Society. Tomorrow's Social History: Classes, Conflicts and Culture in the Programmed Society*, trans. L.F.X. Mayhew. New York: Random House.

Tranfield, D., *et al.* (2000). Organisational learning – it's just routine! *Management Decision*, 37 (4): 253–60.

Tranter, D. (2000). Evolving to reflect the modern industrial life-science environment. *Pharmaceutical Science and Technology Today*, 3 (12): 399–400.

Traweek, S. (1988). *Beamtimes and Lifetimes: the World of High Energy Physicists*. Cambridge and London: Harvard University Press.

Tsoukas, H. and Vladimirou, E. (2001). What is organizational knowledge? *Journal of Management Studies*, 38 (7): 973–93.

Tsoukas, H. (2003). Do we really understand tacit knowledge? In M. Easterby-Smith and M.A. Lyles (2003). *Handbook of Organization Learning and Knowledge Management*. Oxford and Malden: Blackwell, pp. 410–27.

Turner, V. (1969). *The Ritual Process: Structure and Anti-Structure*. London: Routledge & Kegan Paul.

Turner, V. (1982). *From Ritual to Theatre: The Human Seriousness of Play*. New York: PAJ Publications.

Tempest, S. and Starkey, K. (2004). The effects of liminality on individual and organizational learning. *Organization Studies*, 25 (4): 507–27.

Valentine, J. (2000). Information technology, ideology and governmentality. *Theory, Culture & Society*, 17 (2): 21–43.

Virilio, P. (1994). *The Vision Machine*. Bloomington and Indianapolis: Indiana University Press.

Virilio, P. and Lotringer, S. (1997). *Pure War*. New York: Semiotext[e].

Von Hippel, E. (1998). Economics of product development by users: the impact of 'sticky' local information. *Management Science*, 44 (5): 629–44.

Vrettos, N.M. (1999). Unleashing managerial advantage in pharmaceutical research and development. *Drugs Made in Germany*, 42 (1): 13–19.

Wærn, R. (2001). *Gert Wingårdh, architect.* Birkhäuser: Publishers for Architecture.

Wajcman, J. and Martin, B. (2002). Narratives of identity in modern management: the corrosion of gender difference? *Sociology*, 36 (4): 985–1002.

Wallach, M.A. and Kogan, N. (1965). *Modes of Thinking in Young Children: A Study of Creativity–Intelligence Distinction.* New York: Holt, Rinehart & Winston.

Weber, M. (1948). Science as a vocation. In H.H. Gerth and C.W. Mills (eds), *From Max Weber: Essays in Sociology*, London: Routledge & Kegan Paul, pp. 129–56.

Weber, M. (1992). *The Protestant Ethic and the Spirit of Capitalism.* London: Routledge.

Wehner, L., Csikszentmihalyi, M. and Magyari-Beck, I. (1991). Current approaches used in studying creativity: an exploratory investigation. *Creativity Research Journal*, 4 (3): 261–71.

Weick, K.E. (1979). *The Social Psychology of Organizing*, 2nd edn. New York: McGraw-Hill.

Weick, K.E. (1995). *Sensemaking in Organizations.* London: Sage.

Weick, K.E. (1996). Drop your tools: an allegory for organizational studies. *Administrative Science Quarterly*, 40: 301–13.

Weick, K.E. and Roberts, K.H. (1996). Collective mind in organizations: heedful interrelating on flight decks. In M.D. Cohen and L.S. Sproull (eds), *Organizational Learning.* London: Sage.

Weick, K.E. and Westley, F. (1999). Organizational learning: affirming an oxymoron. In S.R. Clegg, C. Hardy and W.R. Nord (eds), *Managing Organizations.* London: Sage.

Weierter, S.J.M. (2001). The organization of charisma: promoting, creating, and idealizing self. *Organization Studies*, 22 (1): 91–115.

West, M.A. and Farr, J.L. (1990). *Innovation and Creativity at Work: Psychological and Organizational Strategies.* Chichester: John Wiley.

West, M.A. and Rickards T. (1999). Innovation. In M.A. Runco and S.R. Pritzker (eds), *Encyclopaedia of Creativity.* London: Academic Press, pp. 45–55.

Whitehead, A.N. (1925/1967). *Science and the Modern World.* New York: The Free Press.

Whitehead, A.N. (1927/1978). *Process and Reality.* New York: Free Press.

Whitehead, A.N. (1933/1967). *Adventures of Ideas.* New York: Free Press.

Whitehead, A.N. (1938/1968). *Modes of Thought.* New York: Free Press.

Williams W.M. and Yang, L.T. (1999). Organizational creativity. In R.J. Sternberg (ed.), *Handbook of Creativity.* Cambridge: Cambridge University Press, pp. 373–91.

Williamson, B. (2001). Creativity, the corporate curriculum and the future: a case study. *Futures*, 33: 541–55.

Willmott, H. (1998). Towards a New Ethics? Contributions of Poststructuralism and Posthumanism. In M. Parker (ed.), *Ethics and Organization.* London, Thousand Oaks and New Delhi: Sage.

Wilson, T.D. and Schooler, J.W. (1991). Thinking too much: introspection can reduce the quality of preference and decision making. *Journal of Personality and Social Psychology*, 60: 181–92.

Winnicott, D.W. (1971). *Playing and Reality.* London: Tavistock.

Winter, S. (2000). The satisficing principle in capability learning. *Strategic Management Journal*, 21: 981–96.

Woicehyn, J. (2000). Technology adoption: organizational learning in oil firms. *Organization Studies*, 21 (6): 1095–1118.

Wolcott, H.F. (1995). *The Art of Fieldwork*. AltaMira, CA: Walnut Creek.

Wood, M. (2002). Mind the gap? A processual reconsideration of organizational knowledge. *Organization*, 9 (1): 151–71.

Woodman, R.W., Sawyer, J.E. and Griffin, R.W. (1993). Toward a theory of organizational creativity. *Academy of Management Review*, 18 (2): 293–321.

Yeoh, P.-L. and Roth, K. (1999). An empirical analysis of sustained advantage in the U.S. pharmeceutical industry: Impact of firm resources and capabilities. *Strategic Management Journal*, 20 (7): 637–53.

Young, A.P. (1999). Rule breaking and a new opportunistic managerialism. *Management Decision*, 37 (7): 582–8.

Young, R.J.C. (2001). Colonialism and the desiring machine. In Castle, G. (Ed.), *Postcolonial Discourses: An Anthology*, Oxford: Blackwell.

Young, R.M. (1994). *Mental Space*. UK: Process Press.

Zaleznik, A. (1977). Managers and leaders: are they different? *Harvard Business Review*, 55 (5): 67–80.

Zhou, J. (1998). Feedback valence, feedback style, task autonomy, and achievement orientation: interactive effects on creative performance. *Journal of Applied Psychology*, 83: 261–76.

Zivin, J.A. (2000). Understanding clinical trials. *Scientific American*, April: 49–55.

Zollo, M. and Winter, S. (2003). Deliberate learning and the evolution of dynamic capabilities. In M. Easterby-Smith and M.A. Lyles (eds), *Handbook of Organization Learning and Knowledge Management*. Oxford and Malden: Blackwell, pp. 601–22.

Index